FARM
FOR LIFE

TANGAROA
WALKER

FARM FOR LIFE

Mahi, mana and
life on the land

TANGAROA WALKER

PENGUIN

UK | USA | Canada | Ireland | Australia
India | New Zealand | South Africa | China

Penguin is an imprint of the Penguin Random House group of companies,
whose addresses can be found at global.penguinrandomhouse.com.

Penguin
Random House
New Zealand

First published by Penguin Random House New Zealand, 2021

10 9 8 7 6 5 4 3 2 1

Design by Cat Taylor © Penguin Random House New Zealand
Cover photograph by James Jubb
Author photograph by James Jubb
Prepress by Image Centre Group
Printed and bound in Australia by Griffin Press, an Accredited ISO AS/NZS 14001
Environmental Management Systems Printer

A catalogue record for this book is available from the National Library of New Zealand.

ISBN 978-0-14-377570-6
eISBN 978-0-14-377571-3

penguin.co.nz

CONTENTS

FARMERS ARE WEIRDOS

One morning at about 4.30, I was at the shed doing the milking. I was usually there by myself, so I got a bit of a fright when I heard a voice calling out: 'Hey mate!'

I replied, 'Hey mate, you all good? What's up?'

He said, 'There's been a car accident. Someone's hit a cow.'

'Righto, mate, I'll come out now. Who hit the cow?'

'A woman.'

'Is she all right?'

'She's dead.'

What?! I couldn't believe it. This was not how I was expecting my morning to go: 'What do you mean she's dead? Oh my god! Have you rung an ambulance? Have you rung the police?'

'Oh, no, I just thought I'd come down and find the first farmer.'

I was so angry but also very confused. I jumped on my bike and boosted up to the road. I was nervous about what I was going to find there but I knew I had to do whatever I could to help.

When I got there, I saw a cow lying on her side completely out to it. I knew I'd have to kill her if she wasn't already dead as she was very badly injured.

Then I saw a lady leaning up against a car that looked pretty dinged up. Now I was even more confused.

'What's happened? Is an ambulance coming?'

She said, 'I'm a farmer. I just live up the road. I realise that at this time of year accidents happen.' We were just near the end of calving, so it was towards the end of winter, which meant the roads were a bit slippery some mornings.

If she was the one who'd crashed and she'd killed someone, how come she was so calm?

'But where's the lady who died?'

It was her turn to be confused. 'There's no one else here. It's just me.'

I looked over to the fella who came and got me. 'Bro! You said someone had died . . .'

He goes: 'I meant the cow, man. You asked me if she was okay and I told you she was dead.'

'I don't care about the cow! I thought you meant the driver was dead.'

I worked out that the heifer had calved then lost the plot because we'd separated her from the herd. She was trying to get back to them, and she managed to clear about four five-wire fences before breaking through our boundary fence and onto the road.

The lady just happened to be driving past on her way to work when the cow bolted onto the road and collided with the car. The driver was really shaken up and her car was a write-off. I was so relieved that she wasn't badly injured and she didn't have her kids in the car. It could have been so much worse.

That whole experience made me realise something. While we farmers can be seen as weirdos, a lot of people out there don't understand us at all. That guy really thought I cared more about the cow than the person who hit her. It made me even more determined to try to share what I love about my job with other people. I want them to understand why we do what we do, and to maybe think that it might be a career that could work for them.

To want to get up at four in the morning every day, you've got to be missing something. Actually, that part's easy once the gumboots are on, but it's that 30 metres from the bed to the boots that you've got to get through. That's the bit that takes that little bit of weirdo to get through each day.

The best thing about the job is that it's forever changing. You never know what you're going to walk into every day. You turn on the lights on your motorbike and your day can change instantly. The surprise aspect of it is awesome. I can go from being knee-deep in mud calving a cow then have to go home, have a shower, get changed and go into town for a business meeting with my accountant or my bank manager.

We're not just cow cockies either — we're running big businesses. Every day, on the property where I work, we milk 550 cows twice a day. The milk this farm produces in a year feeds 3,250,000 people their recommended protein and fat intakes for a single day. So for every hectare (equivalent to two rugby fields) we feed 14,200 people their recommended protein and fat intakes for a single day. That's a quarter of the population of Invercargill.

I've been working on dairy farms since I got my first weekend job when I was 12. Before that, I'd been through some stuff, including going to six schools before I turned six and being whāngaied twice before I was seven. I was six when I went to live with my auntie and uncle on their small farm just out of Tauranga. It was there that I really got to learn about what it's like to work on a farm.

I realise now that that was an experience that heaps of town kids used to have when they'd go and stay at their cousins' house at the farm for the holidays. But now a lot of those town kids never get a chance to go and stay in the country and learn about farming, so it's up to people like me to try to use social media to not only capture people's interest but to educate them without them even really knowing.

That's one of the reasons why I set up my Facebook page, Farm 4 Life. On it, I basically talk a lot about what an awesome job dairy farming is, and I also share heaps of information about what happens on the farm, show some of the jobs we do and talk about some of the issues that farmers face.

While those kids are watching my videos, they not only find out more about dairy farming, which is the backbone of New Zealand, but they might also begin to think about whether it's an industry they might want to work in.

Back when I was at school, I was the dumb arse because they didn't have a measure for the things that I was good at. From 2002 until 2007, I thought I was dumb the whole time. That still sits with me. I still think I'm dumb because I can't spell very well. That's my insecurity, but it was created because the measurements they used within the school system only benefited some people. I know there are heaps of kids out there who feel that way too, and I want to tell them that they're not dumb. Not meeting those tight criteria at school doesn't mean you're going to be bad at business, and it doesn't mean you're not going to be successful in life. You can do whatever you want to do, you just have to really put your mind to it and work hard.

One of the other reasons I started Farm 4 Life was to shed a light on what we do as farmers. It's not just about putting cups on cows, it's about patting them as they walk up the lane and looking after them as best we can. I saw this as a good chance to share my passion for dairy farming and to tell the story from grass to glass.

We farmers get hammered by the media sometimes. There's a lot of negative stuff out there because there are a small number of people on farms who aren't pulling their weight and they're the ones who get focused on. Those people are the ones who are letting us down and they need to be booted out of the industry because they're giving all of us who are trying to do good a bad name.

If you go into any industry, you'll find people who are stepping over the line in certain places. If you take one teaspoon of mud and put it in a bucket of clean water, all that water turns dirty. That's sort of what's happening with us farmers at the moment. There are heaps of us who not only meet all the rules and regulations but go well above them, and those bad ones are the mud that gets chucked in the bucket with us.

When we get smashed with all this negative stuff in the media and on social media, what you don't see are the humans out there. You see all the cows standing in mud, you see the cows in the paddock getting snowed on, but you never see the farmers out there working in this weather, you never see the farmers out there standing with the cows, you never see the farmers up all hours of the day and night saving cows and calves. You never see me crawling out of bed at 2 am, grabbing my torch, leaving my warm house so I can go and save a cow or a calf. What better way to counter it than show the good stuff we do every single day and by highlighting the professionalism within the industry?

More than that though, Farm 4 Life has become a bit of a hub for farmers to come and talk about things that they're experiencing and to ask for help and advice. I'm stoked to be able to help out as many of them as I can, and that has helped create a bit of a community of farmers who might not otherwise have people to yarn to about what's happening on the farm.

In this book, as well as telling my story, I've included some advice about things I've been through and lessons I've learned along the way. It's all in here so you don't have to go through some of those things yourself. While some of the advice might seem a bit like it's all about life with cows, it's not. Most of it can be applied to people in whatever job or industry they might be in. After all, we're all trying to make the most out of our lives and doing our best to get along, right?

1.

WHĀNAU

I was born in Tauranga Hospital. My birth mother's name was Nancy and my birth dad was Hanuera, but we just called him Hanu. At the time, Hanu was the president of the Greazy Dogs, a local gang who are my family. Nancy and Hanu were still together when they had me, but at day two I was adopted by a lady called Aroha, who is my mum. Nancy and Mum were family friends, and because Mum couldn't have children, Nancy gave me to her and her boyfriend Gypsy, who was Rarotongan.

Mum was quite short — about 155 centimetres tall — but she was a very strong lady. She had a little bit of a puku on her and she had really strong calves that people gave her a hard time about. Honestly, they were like little rugby balls attached to a pair of feet!

She was the loveliest lady you could ever meet. I think that's where I get my whanaungatanga from — I always really welcome people into my house because that's what Mum was like. She was the manaaki queen — everyone felt welcome in her home.

My mum used to live with my nana in a place called Welcome Bay. It's a beautiful place on the eastern side of Tauranga moana where the Waitao River flows watched over by the maunga Kopukairoa. Being right by the water meant there was always a supply of kai moana for our marae.

My mum's mum — Nana Mōkai Matthews — was about 54 when I was born. Her husband, my koro, had died about four years earlier. Nana and Koro had built their house themselves and it was right opposite our marae, Te Whetū-o-Te-Rangi. It was a single-storey four-bedroom house with a big yard and a garage where they used to hold heaps of parties. At the back of the yard there were five rows of grapevines, a tree with big, beautiful, juicy golden peaches, a couple of pear trees and some fig trees. There was also a big rokuata tree.

We stayed there for a wee while and I had my first birthday there. Living with the Greazy Dogs meant there was a big hākari or feast for my birthday. They put down a hāngī and there was a big party. My mum was a bit of an alcoholic and she liked to party 24/7. I don't think she had a job then — she just looked after me.

Nana Matthews was a really pretty old lady, who used to drive a Mercedes Benz. It was real styley with a leather interior. It was a flash car back then. She was also a really good cook who was famous for her rēwena bread. She had five daughters and one son — my mum, Aroha, was the eldest and then there was Denise, Peata, Moana, Jan and Uncle Alby.

My great-grandmother, Nana Matariki, lived next door to

Nana's house, and Mum used to take me to visit her all the time. Eventually, she moved into an old people's home, so I didn't see her as often after that.

I was about three when both Nana Matariki and Nana Matthews passed away. Mum was the oldest of six children, so she thought that she would get the house. But one of my aunties moved into the house instead and we got kicked out. That's how we ended up living in a tent out the back of her section. My mum's brother, Uncle Alby, used to live in an old run-down van that was parked opposite us. The tent was under a big pear tree surrounded by really long grass next to the fruit orchard. Having fresh fruit was awesome, but it also meant there used to be heaps of fruit flies everywhere. They hung out on the rotting fruit and got on everything.

The fruit trees also attracted heaps of wild ducks, which me and my cousins used to try to catch. We never had much luck, but we did get a few eggs off them!

Obviously, living in a tent, we didn't have a bathroom. We had to wash ourselves in a big plastic container with handles on the side like you'd put your washing in. That used to be our bath. Mum used to fill it up with water, then bath me in it, then she would wash herself with a bad old tea towel.

That container got used a lot. Mum would wash all our clothes in it and use it for doing the dishes. She also used it for cooking — mixing up rēwena bread, making the pastry for apple pie, you name it.

Mum was a vicious apple-pie maker. She was probably the best apple-pie maker in the whole of New Zealand, I reckon. She was a bit of a legend at it! She was a really good cook and she showed her mana through food. We never had the flashest things, but if there was a party on she'd turn it on. She used to cook everything, including her apple pies, on a little barbecue.

My auntie used to be in a band, and she had all her bandmates living in Nana's house. It was really weird that they lived inside but we had to live in a tent at the back of the section. I still don't understand how anyone could do that to their sister, let alone their sister who had a little fella. I've never liked the way that all rolled out.

We lived there for a few months and were never allowed in the house. Sometimes when my auntie was away at her gigs, we'd sneak into the house and have a cheeky bath or a shower. That would have been the first time I used a toilet, but I guess I was lucky that most of the time we were living in the tent I was still in nappies.

Eventually, my auntie moved to Australia and we haven't seen her since. After she left, me and my mum got to live in Nana and Koro's house. As well as us, there were quite a few of my cousins living there. By this time, I was about three or four, and I remember all the drugs and alcohol, and Mum being with random men.

About the time I was born, my mum's sister, Peata, and her husband, Kelly, came back from Australia after living there for a long time. Kelly was from Te Puna on the other side of Tauranga Harbour. They lived in a place called Whakamārama where they had 40 hectares of land. It was a way up in the bush, and the place was covered in gorse and ragwort when they bought it. They didn't have the money to build a house, so they got the builders to build a double garage and put ranch sliders in instead of a garage door.

Eventually, Uncle got some friends to insulate the place, and it's grown with them over the years, but when I first went there it was very basic.

When Mum would go out on the piss with her mates, Auntie Peata and Uncle Kelly would come and look after me. The story goes that before my nana died, she said to Auntie Peata, 'That's your boy, Peata — you have him.' I ended up going to stay with them pretty much every weekend.

One night, when I was about four, my mum went out on the piss and left me at home. My cousins, who lived right next door, were looking after me. They used to boil up sugar and whatever else went in the pot to make toffee. I loved eating that stuff. It's probably why I've got no teeth now!

I got four chairs and chucked a sheet over them and made a fort. I crawled in there and waited for my cousins to go to sleep. As soon as they'd all crashed out, I got up and went to the kitchen. I packed all the toffee into a little lunch box and put it in my backpack. Then I went and got my little purple and green puffer jacket off the clothes horse and put it on. Next, I put my gumboots and backpack on and jumped on my trike.

With all that done, I rode off down the road in the dark trying to find my mum. It must have been about one in the morning. I thought she might be at the marae, which was about 300 metres down the road, so I headed over there. I could see the marae because there were lights on, but I was really scared being out at that time of night on my own.

When I got down to the marae, I found heaps of people there, including some of my aunties, and they were all on the piss. I went up and asked where Mum was. Everyone was like, 'What are you doing here? Where's your mum?' She obviously wasn't there. One of my aunties ended up walking me back to the homestead and leaving me with my cousins.

The next morning, Mum arrived and was banging on the door. It was still dark so it must have been pretty early. When she came in, her whole face was covered in blood. She was bleeding from the head all the way down her body and she had all this glass in her head. It turned out she'd crashed her car and she'd gone through the windscreen and then been pulled back into the car. She was cut up really badly.

It gave me a real scare and I cried hard out when I saw her. Mum told me to run outside and grab the big plastic container that we used to bath in. I grabbed it and put warm water in it, then she told me to put some salt in it.

Mum sat on the couch and leaned over the salty water, still bleeding everywhere. I was on my knees in front of her, and I put a towel over her head to help clean up all the blood. She was in a really bad way. I know now that she was still drunk when she got home that morning and that she had driven drunk and crashed her car.

Just as it was coming to daylight, Mum comaed out on the couch with the towel over her. I was just looking after her as well as I could and crying. I didn't know what else to do so I decided to ring my auntie and uncle. Thankfully, our old cord phone had their number programmed in, so I knew which square to push to get through to them.

I told Auntie Peata what had happened and she and Uncle Kelly came around straight away and picked us up. They dropped Mum off at hospital and took me up to their place at

Whakamārama, which is about 20 kilometres out of Tauranga on the road to Katikati.

From then, it was a big decision for my mum to wake up and realise that what was happening wasn't very good. At the same time, my auntie and uncle thought, 'Maybe we should adopt this boy and take him under our wing . . .'

I ended up staying with Mum. She still wanted me, and she didn't want to give me up. She wanted to try to change the way she was. What my auntie and uncle didn't know was that I'd seen Mum slicing her wrists in bed at night. She slept with a knife under her pillow and when things got a bit much for her, she'd cut herself. She had marks all over her arms, but I was too young to understand what was going on.

A few times, she told me that she hated life and she wanted to kill herself. Then she'd cry and thank me for being there. I used to just lie there and cry with her, saying, 'I don't want you to die, Mum!' She'd say, 'You're the reason I'm still here.' Then she'd thank me for being there as I gave her a reason to stay alive.

It was about this time that my auntie sold my nana's house, so we had to move out. It was really sad because my nana and koro had built that homestead. I loved that place so much. Before we left, I stood outside the house with my cousin Taipiri. 'One day, bro, when I get rich, we'll buy this house back!' I told him. I've never forgotten that promise and one day I will buy Nana's house back.

Mum and me moved out of there and into a little shed just down the road. It was just below my Auntie Manu's house. It was cold and so damp that I had to clean the mould off the walls. There were heaps of rats and mice, and heaps of cockroaches. We used to catch the cockroaches in old cassette cases. Once we'd got about four or five in the tape case, we'd chuck them on the fire and watch them burn.

The shed was right next to the river, so it used to flood quite often. When it did, the driveway would be covered with water, so we couldn't leave.

When I used to go and stay with my blood family, there was just a tonne of dope and alcohol around. It was a normal thing for me to sit there and watch my aunties and uncles having a great time just getting really out of it. We used to sleep in the lounge marae-style with us all on mattresses on the floor. I loved lying there with all my cousins watching TV, while all my aunties and uncles were there pissing up, smoking dope, playing the guitar and having a good time. There was just so much love around. That used to happen most nights and it didn't matter what day of the week it was, but they always used to get up in the morning and go to mahi.

The family ran a kiwifruit orchard, and it seemed like all of the aunties, uncles and cousins worked there. When they were at work, me and my cousins were free to do what we wanted, free to roam, but we knew we were loved. My cousins, who were only a bit older than me, used to drive these cars around the orchards and were always crashing them. I guess we grew up really young. Some of us were driving cars before we even went to school.

We'd go down to the estuary and get kai moana like tītiko, and we'd go floundering to feed all the whānau. We lived really close to the marae, so whenever someone would die or get sick, we'd go down to drop off kai and look after whoever needed help.

All through my childhood, I used to go to visit my blood brothers and sisters, and my blood mum and dad, Nancy and

Hanu, on the weekends. The kids were all drinking at a very young age. I was five when me and my cousin Ryan stole a packet of cigarettes from one of my uncles while he was on the piss. We went behind the tank at my nana's house and rolled up a cigarette and smoked it. I absolutely hated it and haven't had a cigarette since.

My uncles there were in the Greazy Dogs, and my cousins had their own gang out there. They all used to hang out, do breakdancing, listen to Tupac and think they were little gangsters. There was quite a bit of violence then, too, and I used to often see them having fights.

While we were living in the shed, I turned five and started school. I went to Otepou School, which was a Māori school in Welcome Bay. I must have been there for about four or five weeks when a kid called Tama stole my pen off my desk. I grabbed it off him and stabbed him in the throat with it. I don't know why I did it, I guess I was quite an angry kid. The teacher grabbed me, slapped my hand, then sent me to the principal. I got expelled.

Mum was really angry at me for getting expelled, but also because she knew that going to a new school meant I wouldn't have my cousins to look after me. I think she was also pretty pissed off that I got whacked by my teacher.

Mum enrolled me at Welcome Bay School, but I ended up being there for an even shorter time than I was at Otepou. Mum hadn't been working, and she didn't like living where she couldn't get out whenever it rained, so she decided it was time to leave. Not long after, she got work cleaning houses and old people's homes in Mount Maunganui and we moved to a little flat on Girven Road.

Two of Mum's cleaning jobs were out in Pāpāmoa, and we

didn't have a car, so we used to bike out there. I regularly used to bike all the way out there from Bayfair following my mum. It was a long way for a five-year-old on a little pushbike. It would take us about 40 minutes to bike out there, then Mum would clean these two houses and we would bike all the way back again.

Before long, I enrolled at Papamoa Primary School, which was about 10 kilometres from our flat. Me and Mum would bike out there, she would drop me off at school, then go to work. After school, she'd bike back down and pick me up and we'd ride home together — rain, hail or shine.

It wasn't long before Mum got fired from her job because some old man groped her, and she slapped him. That meant we couldn't afford to stay in the Mount, so we ended up having to move to a place in Tauranga. Mum's partner Gypsy moved out there with us, so there were three of us living there. Gypsy had come to New Zealand from Rarotonga in the 1970s. He met my mum down in Wellington, and they were together on and off from then on.

Gypsy was always on the scene, but he wasn't always around. He'd come and go, and Mum would see other men. A lot of the time he was either away working as a fisherman or he was in jail, so he'd go away for a long time then come back and they'd get back together. He always supported Mum when he was around, but he was a bit of an alcoholic too. I called him Dad, but that was just his name. He wasn't really a father figure to me.

Gypsy was a pretty tall, well-built guy, but Mum knew how to get the better of him. One day, I walked out of the house to see my mum smashing the windscreen of Gypsy's car with a sledgehammer. He'd obviously been mucking around with someone and she wasn't happy about it. It wasn't long before we left that house and moved out to Te Puke. Back then, Te Puke was a beautiful little town where people used to stop off on the way to the eastern Bay. The main attraction was a wicked bakery there where you could get 12 mussels and some fish and chips for $10.

When we moved out there, we lived in a caravan park. I thought it was pretty cool because there was a swimming pool and a tennis court. A lot of people who lived there worked at the local kiwifruit packhouses, so there were lots of other kids there, and I soon had lots of mates. The kids in Te Puke were very mature for their age. They were all in gangs and they knew all about stand-over tactics. They all knew how to fight, and they loved a bit of biff even when they were quite young. They were haati kids, tough as, and you wouldn't muck around with them.

The local hangout was the Te Puke Citizens' Club and I used to go down there quite often with Gypsy and Mum. They'd go and play on the pokies and I'd sit and eat a seafood basket. Other than that, my favourite place was the local swimming pool.

When we were staying at the caravan park, Uncle Kelly would ring Mum and tell her when he was coming through in the cattle truck he was driving at the time.

I'd be in the caravan and I'd hear this 'Toot-toot-t-toot-toot!' I'd run out to the road, and Uncle would be parked there waiting for me.

He'd say, 'Jump in, my mate!' and off we'd go on an adventure.

The first time he picked me up, he took me to the Rangiuru cattle saleyards out by Paengaroa. That was probably my first introduction to farming.

While we were in Te Puke, Mum worked in a kiwifruit packhouse and was doing some house cleaning, but she was still drinking all the time. It wasn't long before she got behind on the rent on the caravan, so I got pulled out of Te Puke Primary School and we moved all the way out to Waihi Beach about 80 kilometres away.

Mum's cousin had a place right on the beach there. We rented the bottom part of her house and she lived upstairs. It was beautiful living right by the sea. We always had heaps of seafood and it was cool to be able to go and play at the beach. I went to primary school while we were living in Waihi Beach.

Mum was working quite a bit so I just got left to do what I liked. One time, I'd gone up to the shop and I was biking back down the hill to the house and I hit a pipe that was across the road. I went straight over my handlebars and grazed my chin all the way down the road. It was grazed right down to the bone. My chin was so mangled, it was a real mess.

When I got home, Mum filled up a margarine container with warm water and salt, so I could bathe my chin. There was blood everywhere. It was a Friday afternoon and she didn't want to let it stop her going out partying, so she dropped me off out at Auntie Peata's and Uncle Kelly's place in Whakamārama. When they saw my face, they took me straight to the hospital. There, the doctor told me that the only thing that had saved me was the buckle on my helmet. It had been ground down to nothing as my face was dragged along the ground.

I spent the weekend at my auntie and uncle's, and Mum finally came to get me after a weekend of partying with her mates. As well as being a heavy drinker, Mum had quite bad diabetes that meant she needed to take insulin to keep her blood sugar levels stable. Of course, that didn't really mix well with the alcohol, so her health was always a bit sketchy.

Not long after that happened, we moved back to the caravan park in Te Puke. We'd only been there for a few weeks when a teacher from the local high school, Malise August, came into our lives.

Malise had a room to rent in her small two-bedroom house. It was quite cheap, so we moved in. She seemed to have everything sorted out. She was really well presented and very proper. She was studying and she was working a full-time job as a teacher. She had her own place, a nice car and a stable relationship. She was just really together. I didn't have many people around me like that, so she made a real impression on me. Looking back on it now, I can see that she played a bigger role in my life than I ever realised. Her life was so different from everything I knew back then, and I guess it made me realise that things could be different.

Malise was a big influence on my mum as well. While we were living there, Mum went from not having a job to looking after old people. As well as working, she enrolled to study for a certificate in computing. With some of the money she earned, she even bought a computer. Back then, they were still pretty expensive and not everyone had them. It was so cool.

Before long, she had a job at a kiwifruit company doing computer stuff, which meant she had enough money to buy a car. Life was pretty good for a while, but then Mum's health took a turn for the worse and she had a stroke and was admitted to hospital. She was in her late forties. Things really started to go downhill from there.

Up until this point, I'd been spending the week with Mum wherever she happened to be living, then in the weekends I'd be with either Mum's family or my birth mum's family at Welcome Bay, with all the drinking and pot smoking, or with my auntie and uncle in Whakamārama.

When Mum got sick, I had no one around to look after me. When she started drinking again she had no money and her health suffered even more, so Auntie Peata and Uncle Kelly decided that I should go and live with them. Their own kids were old enough to be my parents and had long since left home. Auntie Peata's son, Jarrod, lived up north and Uncle Kelly's two kids, Nicky and Grant, lived in Melbourne.

When I moved in with my auntie and uncle, it was the sixteenth house I'd lived in and Whakamārama School was my sixth primary school — and I was only six. All the other schools I went to had heaps of kids at them, but I never made any friends because I was never there long enough to build relationships with other kids. That turned me into quite a shy kid. I'd just go into my shell and not bother to try, as I knew I'd be leaving soon. I also had all my cousins around me at the weekends, so I didn't really feel like I needed friends. The main thing I remember about going to most of those schools are the lunch orders!

Whakamārama ended up being my forever home. Uncle Kelly was a real dag. My uncle's an out-of-the-gate loose guy — he just tells people how it is and doesn't care what they think. If someone needs to be told, he'll tell them. If he doesn't like you and you go to shake his hand, he'll just stand there and look at your hand. He's bloody straight-up, so I always knew where I stood with him and I fully respected him.

While Uncle was a real cheeky bugger, Auntie was, well, a bit proper. She was a feisty lady and a really hard worker. As well as working long hours, she spent a lot of time looking after the garden and doing the lawns. Even so, she always looked nice and was well presented.

BRINGING MĀORI PRINCIPLES TO OUR MAHI

From when I was really little, my mum taught me about the importance of manaakitanga. She was the manaaki queen and it's something I've really tried to bring into my work life.

Manaakitanga is about caring for those around us. Kaitiakitanga is about protecting our planet and people. Whanaungatanga is about creating lasting relationships with those we bring into our workplace.

Some people have asked me why I feed people who are working on the farm with me. I tell them it's just what you do. For me, manaakitanga focuses on positive human behaviour and encourages people to rise above their personal attitudes and feelings towards others. The aim is to nurture relationships and to look after your guests no matter what level of support they bring to your business.

2.

LUNCH MONEY

At Whakamārama School, there were 63 pupils. My first teacher was Lyn Harrison — Miss Harrison — in Room Three.

In about my second week at school, we were learning about other cultures, so this Indian guy came up and brought all these different types of curries for us to try. I smelled one of them and felt really sick straight away. I asked the teacher if I could go to the toilet. She said okay.

When I got there I tried to spew up, but nothing happened. I went back to the classroom, then felt really sick again. I put my hand up. 'Miss, can I go to the toilet again?'

'No, you've just been.'

'But I'm feeling really sick, Miss.'

'You've just been. Sit down.'

I didn't have any choice but to sit there feeling sick. I was sitting with this kid, Blair, who lived up the road from me. He said, 'Let's just go, bro!'

The pair of us got up and went into the toilets. I felt really sick and I was a bit short of breath, which made Blair a bit scared. He ran up to the principal's office and she rang my uncle.

Uncle Kelly rushed to the school and picked me up. He took me home and I went straight to bed.

Uncle was in the kitchen and I yelled out to him, 'I can't breathe!'

He drove me down to the local doctor's surgery. By the time we got there, I had these red spots all over my skin. Straight away, he told my uncle that he thought I had meningitis and then he rang an ambulance. All I really remember was him flipping me over and sticking this big needle in my arse, which knocked me out.

When we got to the hospital, it was like I was in a movie. I was lying on my back as they zoomed me through the corridors. I watched as the ceiling lights raced past. Eventually, I got hooked up to a heap of monitors and they did a whole lot of tests on me.

I was so sick that I ended up being in intensive care for more than a week. I was knocked out for most of the time. When I finally woke up, my uncle was there holding my hand and crying. I lay there with my eyes shut for a while and listened to him talking to me. Then I just went 'BOO!'

He cracked up laughing. 'Bloody hell, my mate, I thought I'd lost you . . .'

It was a long road to recovery for me. I'd lost a lot of weight and they had to give me heaps of steroids to try to get my immune system sorted. For the next month, Auntie and Uncle had to drive me to the hospital every couple of days to get blood tests done, and I put on a load of weight because of the steroids. It turned out I'd had meningococcal meningitis, which is a bacterial infection of the membranes that cover your brain. It's such a fast-moving disease that the doctors reckoned that if I'd got there any later I might not have pulled through.

It took months before I finally got back to normal. That kid, Blair Mason, who'd run and got the principal, ended up becoming one of my best mates. Me, him and Patrick Whittle soon started doing everything together, and we're still best mates now.

Patrick's mum owned the Kip McGrath tutoring centre in Tauranga, and Patrick was a very intelligent boy. Blair's parents owned a framing factory and his dad had a roofing company. My two best mates had parents who were really successful in business and they had really nice houses and nice cars.

My auntie and uncle had a bit of land and that double-bay garage that they'd put walls on and turned into a house. Uncle drove a concrete truck while Auntie worked for the egg company. They weren't rich, but they worked really hard for what they had.

Through that first year of schooling, I made heaps of friends. I guess I was pretty lucky as I was just a kid who lived next door. No one at the school knew anything about my family in Welcome Bay, so they didn't judge me by them. Some of the kids would ask me why I lived with my auntie and uncle, but other than that they just accepted me as another local kid.

I was a bit ashamed of living with Auntie and Uncle at the start. I was embarrassed about the situation because everyone else at school stayed with their mum and dad, and I'd always be going away back to my real family for the weekends.

At one point, one of the other kids at school said they thought they'd seen me in Te Puke in the weekend with someone who wasn't Auntie or Uncle. I knew they would have seen me with my mum, but I lied and said I had a twin who lived over there and they must have seen him. I didn't want to tell anyone I used to live in Welcome Bay or Te Puke. I thought all of the people around me had everything — stability, nice houses, lunch orders and all that — and I didn't want to feel like I was being judged. It wasn't until much later that I realised that they all had their problems as well.

As well as learning lots at school, I learned heaps about how things worked on the farm. I was that kid that when my uncle would get home from work, I'd run outside to go and help him. I'd pull my gumboots on and run out to catch up just because I wanted to be with him. Wherever he was going, whatever he was doing, I'd want to be there with him. Whether he was setting up fences, starting the pump, feeding the cows, whatever, I was always out there.

It would get to lunch order day and all my mates at school would be ordering pies and cakes and orange drinks. I wanted to have them too, so I said to Uncle, 'Can I have five bucks so I can have a lunch order?'

He said, 'Nah, you've got to work for it! If you go out every day, get that cow in, separate the calf from the cow overnight and put the calf in the yard, then in the morning go back out, get the cow in and milk it, I'll give you five bucks a week.'

Monday to Friday, when I got back from school, I'd take the calf off the mum, put it in the yard, then in the morning I'd help Uncle milk the cow. Once the cow was milked, I'd carry the bucket up over the hill to feed all our pigs. We always had five or six breeding sows and heaps of piglets, so I used to feed them some of the milk from the cow and kiwifruit from the local orchards.

Once the pigs were fed, the remainder of the milk went to our pig dogs — there were about 12 or 13 of them — and the local wild cats. I'd just pour a bit of milk onto the lid of a 20-litre container and the cats would emerge from under the house and out of the paddocks. There'd be heaps of them all in a circle, trying not to touch each other as they competed for the milk.

Some nights it would be dark, and we'd be sitting there eating dinner and Uncle would say, 'You get your cow in?'

If I shook my head, he'd say, 'Well, go on then.'

It didn't matter how much I moaned about doing it, I knew I'd have to go out and get the calf off the cow in the dark if I wanted my lunch order money that week.

Doing that job gave me a sense of ownership of it. It was just something I did, but I didn't relate the money thing as being part of the job. I mentalised it as 'I'll do this for Uncle and Uncle will give me five bucks', rather than 'If I do this job, I'll get five bucks'. It wasn't for another few years that I understood the value of working for money.

We usually had between 25 and 30 cows on the farm, so as well as bringing the milking cow in, I used to go and feed out with Uncle most days when he got home from work. He'd be driving the tractor and I'd be on there with him. It wasn't long before he showed me how to drive the tractor, and he'd jump out and do the feeding out while I was driving. That meant I was pretty familiar with how a tractor worked from a young age.

I used to love going out pig hunting with Uncle. Our main pig dog's name was Devil. He was a Ridgeback/Bull Terrier cross. He was my best dog. He was a beautiful golden dog, really handsome. Other dogs would try to fight him, and he would just destroy them.

Sometimes, I'd grab eggs from the chickens and feed them to him. We had this crazy white rooster that was blind in one eye. When I was little he used to chase me. One day, he came bowling down the paddock towards me. I took off running and, when he caught up with me, the rooster started clawing at the back of my legs. I was scared because this thing was a psycho. Out of nowhere, Devil turned up and — whoompa! — that was the end of that chicken. That dog was so protective. He was awesome. He was almost like a big brother to me.

Every day after school, Blair and Patrick and me would be around at each other's houses. We lived four or five kilometres apart, but we were always on our pushbikes heading to see each other. We spent heaps of time sliding through the paddocks, chucking cow shit or sheep shit at each other or having moss fights down at the creek, which always escalated when someone would get hit in the eye or smashed in the face.

Although we only had 60 kids at our school, Blair, Patrick and me were always super competitive. It didn't matter what it was — swimming, cross country, high jump, shot put, the whole nine yards. We were always trying to beat each other at everything.

When we were walking between each other's places, we even used to have competitions to see who could throw a rock at a road sign first. You weren't allowed to move forward until you'd hit the sign, so we'd stand there for hours until someone hit the sign with a rock. Sometimes we wouldn't hit the bloody sign until it was dark or we'd run out of rocks!

My uncle had this old paddock basher, a Suzuki Jimny. All the boys would come around and we'd jump in this bloody thing and try to learn how to drive in the paddock at seven years old. One day, Blair hit a big bump in the paddock and flew up into the roof and his head left a big dent in it. We all thought it was hilarious!

I sometimes used to go to the egg factory where Auntie Peata worked. I'd spend the day going around all the chicken sheds picking up eggs. Every now and then, when I got a chance, I would pinch a few eggs then go and throw them at trees. That was really fun.

Auntie Peata also taught me how to cook very badly. She was an awesome cook, so the 'badly' part was me. When she cooked, she always tried to perfect anything she made. She'd get on a mission where she'd make the same thing five or six times in a row until she smashed it and it was how she wanted it to be. Unfortunately, for me and my uncle, we'd have to eat the same thing for days in a row. The one that really sticks with me was this crabmeat pasta dish she made. It was really nice — for the first two days. By day five, I was a bit sick of it.

She used to do some really nice desserts, though. She got on a stint where she was trying to make pavlova. I'd say, 'How are you going, Auntie?'

'Ohhh, it's looking good, it's looking good!' she'd reply.

I'd look in the glass in the oven and say, 'Oh yeah! Look how big it is!'

Then she pulled it out and it went from being about six centimetres tall to being about point five of a centimetre. We had a week of eating flat pavlovas before she got it right.

She was a really good horse rider, and eventually she taught me how to ride. She and Uncle belonged to the Katikati Hack and Hunters, and they used to ride with them on Wednesday nights and go on weekend treks all over the Bay of Plenty.

For the first time, I had real stability and reliability in my life, but I was also worried about Mum as she was getting really sick. She was having heaps of strokes and lost all the feeling in the left side of her face. She also had a lot of complications with her diabetes. She was taking about 35 different tablets a day to try to keep the diabetes under control, but some of them caused complications with her strokes.

As a kid, going to the hospital all the time and seeing Mum like that was really scary. She was only in her late forties, but she'd aged heaps. She couldn't walk properly, and she couldn't dress herself. She had to drink out of a straw, and she had trouble eating because of the effect the strokes had on her face.

She wasn't able to work much because she was still so sick. Without Mum working, Gypsy struggled financially, so they lived very much week to week. That meant that they weren't eating very healthy food, which made Mum's condition worse.

Even though they were broke, Mum decided to give me some advice about money. She was serious when she said to me, 'Make sure you marry a rich Pākehā girl.'

She obviously thought that was the way out of the hole. I didn't understand that, and I thought to myself, 'But I'm going to be a rich Māori boy . . .'

As well as worrying about Mum being sick, in the back of my head I had a bit of fear that because I wasn't with her, she might go through with her old threats of killing herself. I didn't know how to talk about it, so I never told anyone how I was feeling.

She had a carer, who looked after her during the week, and I used to go and stay with her most weekends. While I was there, I'd clean her and look after her. Some weekends, Gypsy would take me to the Cut at Maketū and we'd go fishing for kahawai. We relied on that to get our kai.

When I was about seven, Mum moved out of Malise's place because Gypsy wanted to move in with her. They moved into a one-bedroom house in Macloughlin Drive. Gypsy would sleep on a mattress in the lounge and Mum slept in a double bed in the bedroom. When I went to stay with them, I'd always sleep in with my mum. At about that time, Mum's medication started to come right, so she was able to work on her diet a bit more.

Even though she was trying to clean up her diet, as soon as she was feeling a bit better, she got back into the drinking and started pissing up with my aunties again. When I was supposed to be staying with her on the weekends, I ended up going to my cousins' house instead. They were meant to look after me and my cousin Ryan while Mum was out drinking.

Ryan and me were eight and the rest of our cousins were all 14 or 15. We'd go out with our older cousins and break into cars and steal any money we could find. One time, we broke into this van and stole all of the change out of the ashtray, some tools and this guy's lunch. He had Krispie biscuits in his lunchbox and they were my favourite. Our cousins used to give me and Ryan anything under $20, as they saw that as chump change. We thought it was heaps of money! For eight bucks each, we could go to the pools and buy a packet of Krispie biscuits.

All of a sudden, we thought it was okay to break in and steal things. We learned how to use a fork to pop the locks on cars and we used to do it all the time on the weekends. I thought it was great that I could make eight bucks for doing nothing more than bending a fork.

Come Monday, I'd go back to my normal lifestyle at my auntie and uncle's place. They didn't smoke, they didn't drink, and they were real straight-shooters. During the week, I'd bring in the cows every day to earn my five bucks a week, and I'd hang out with my mates at Whakamārama, who were total country kids from really stable, successful families who owned businesses. I was eight and already I had two totally separate lives that were the complete opposite of each other.

3.

TE PUNA MANA

When I was nine, Uncle said to me, 'Do you want to play rugby?'

I was like, 'Oh yeah, okay!'

His family are Pirirākau from Te Puna, which is about halfway between Whakamārama and Tauranga, so he's a blue-and-black Te Puna supporter through and through.

I used to go down and watch the rugby with Uncle quite often, and that's how I got to know all of my uncle's whānau, which I'd been whāngaied into. Through them, I became haati Te Puna as well, playing prop for their schoolboy team.

All of a sudden things started changing for me big time. Taking up sports meant I couldn't go to my mum's every weekend because I had rugby. Mum and Gypsy had a real piece-of-shit car, so it meant a lot to me when they came to watch me play. Sometimes, when they came to watch, I'd go home with them afterwards.

Some weekends, my brothers and cousins would come up to Whakamārama to pick me up. They'd come up in their own crappy car that would overheat coming up Whakamārama hill. We could always hear their loud stereos before we saw them. Then when they pulled into the driveway, we'd hear their car bonnet pop and there'd be this loud hiss followed by a gushing sound as their radiator blew all over the place. When I think about it now, I realise it was a long drive out there from Welcome Bay, especially given they had nothing much. I have so much respect for them that they would come out to pick me up so they could spend time with me.

When I was 10, some of the guys at the rugby club told Uncle that he should put me in for the trials for the Tauranga West team that was going to that year's Tai Mitchell Tournament.

At that tournament, 10 teams of primary school kids from around the Bay of Plenty played for the Henry Taipōrutu Mitchell Shield. It's been going since 1938 and it's a big deal in the region. Tai Mitchell was a community leader, who worked closely with the legendary Māori leader Āpirana Ngata, and did heaps of volunteer work. One of his roles was Secretary of the Bay of Plenty Rugby Football Union, so that's why there's still a tournament named after him.

I went to the trials and ended up making the team and that was my introduction into rep rugby. We had to drive into town to Tauranga Intermediate School, where the team would meet to train and do our fitness work. Because Uncle Kelly was driving the concrete truck and doing stuff on the farm, it was a struggle for him to get all his work done and get me into town to my practices.

Uncle gave me a bit of an ultimatum. He said that if I wanted to play for the team, I had to feed out the cows by myself. That way, when he got back from work, he could just pick me up and take me to training.

From then on, after school I'd head up to the paddock where the tractor was. I'd grab a hay bale and put it on the back of the tractor. Then I'd start the tractor up and drive to the paddock where the cows were. Once I was there, I had to put the pin in the steering wheel so it would be locked, and I'd stick the tractor in gear so it was moving really slowly. Then I'd jump off, go around the back and start feeding out the hay.

For the first few times, the tractor was going too slow and the cows would all get into the hay before I managed to feed it out, so I had to jump back on and put it up another gear to speed it up a bit!

I was only 11 and it was an under-13 age-group team, so I had a lot to learn. They worked us boys really hard, and it was the first time I learned about fitness and what it did to the body. I'd never been coached like that before. I was taught about using my aggression and the importance of having a 'don't give up' mentality.

I was a fat kid playing as a prop, but to play in the tournament we all had to be under 53 kilos. To make weight, some of us did some pretty extreme things. One of the bros, Reggie Manu, used to go out running in the middle of the night. I was on the borderline as well, so I wasn't eating much. Some fellas cut their hair off to lose an extra few grams.

That year, the tournament was in Te Kaha. We were staying at a Te Whānau-ā-Apanui marae. When we went to get weighed in at the start of it, some of the players got weighed in the nude to make the 53-kilo cut-off. There were some really grumpy, hungry kids there that day. Once we all made the weigh-in, we all went and smashed heaps of food. It was wicked.

The Ōpōtiki team won the shield and I think we got third. Not winning didn't worry me. Just having the experience of going away to a week-long competition with a team of fellas I didn't go to school with, and who I didn't really know, and playing rugby every day was out the gate!

Blair was also a very good sportsman, so he was in the top local soccer, cricket and basketball teams. We were both really proud to represent our little school up in Whakamārama. I was pretty proud to represent Te Puna, too.

Te Puna was so different from other areas around Tauranga. For some reason, we hadn't been influenced by the drugs and the drinking so much. There were a lot of successful, clever people about there. I reckon it's because there were heaps of marae out there and everyone grew up as marae kids. I also think a lot of the Te Puna mana came from their rugby team.

Uncle Kelly was coaching the Te Puna Colts team and I used to be their ball boy. I really looked up to the Colts and Seniors players back then. They were the cool guys with the cool clothes and the cool cars. They were like the Jonah Lomus of Tauranga! I'd watch my uncle coaching them and he'd send these really cool guys home crying if they didn't make the team. He'd tell them to toughen up and they'd call their mums. Later on, when we were at home, the phone would go.

'Kia ora, this is Tangaroa.'

Down the phone would come an angry female voice. 'Where's Kelly?'

I'd go and get Uncle, then I'd wait to listen to the conversation. Usually, the mums would be shouting so loud, I could even hear them.

'How dare you be growling my son! He's playing recreational rugby and how dare you be telling him he's fat and unfit!'

Uncle would tell them what he'd told their sons: 'Toughen up! I tell it exactly like it is, and if you don't like it you know what you can do . . .'

There was a really good culture at Te Puna. There was such a difference between the role models I had there and the ones at Welcome Bay, where my cousins were all in gangs and out breaking into cars. The biggest difference was the way they held themselves. My cousins from Te Puna were really proud, confident Māori, who were staunch in where they were from and their whakapapa.

My cousins at Welcome Bay were also proud and staunch, but I don't think they were as confident in themselves, or in where they were from. I reckon they were trying to find themselves.

Both lots of cousins were trying to work out who they were and to prove their mana to those around them somehow. The paths they took reflected what was around them and what they'd seen growing up.

Coming from Te Puna, my cousins could prove their mana among their peers on the rugby field every weekend from the age of nine right until they were in their thirties. Whereas, when you're in a gang, you have to prove your mana by having fights, telling people to f— off, having a gun, having a knife, beating someone up, doing outrageous illegal things, selling drugs and all that stuff. In a way, they were all trying to do the same thing but in very different ways. We were all in a big struggle to find ourselves and they were all just reacting to the world that surrounded them.

My cousins in Welcome Bay, their parents didn't always have their backs. Their parents were always out on the piss and they were left to their own devices. They'd go out and make their own mischief as a result. No one made them go to school — in fact, they got told that only dummies went to secondary school. Everyone was just doing what they had to do. When my cousins in Welcome Bay got a growling, their parents would tell them to shut up and really yell at them. As a young kid there, you would be ruled by fear. If you got kicked out of school, the reaction would be, 'Oh well . . .' If you got sent to jail, 'Oh well . . .' If you got up the duff when you were 16, 'Oh well . . .' Things were just accepted.

In Te Puna, the parents expected more from their kids. For a start, kids would never be sworn at by their elders. All it took was one look and the kids knew they were in trouble. Rather than being ruled by fear, they were ruled by respect and mana. They wouldn't want to disrespect their parents, aunties or uncles. Out there, kids would see their parents get up on the marae and have a kōrero. They were teachers at the kura or they were kapa haka teachers. They were always community people and they were very proud of that. They really valued education, so much so that a lot of their kids got sent to boarding schools in Auckland.

While I wasn't headed for boarding school like some of the more well-off kids, my auntie and uncle could see that it was a good idea to keep me busy. Right from a young age, they got me into heaps of sports. In an average week, I would go to swimming training in our school pool on a Monday, rugby training then basketball practice on Tuesday, on Wednesday was swimming again, on Thursday rugby practice, and on Friday night I had two or three games of Miniball. On Saturday, I had two or three games of rep basketball, which was followed by a game of club rugby. Then I played rep rugby on Sunday. I was doing something every single day and I don't know how my auntie and uncle kept up with it. They managed to get me to every training and every game on time.

As well as playing rep rugby, I had made a rep basketball team and I was selected for a regional swim meet. Ella Kirkham used to take me for swimming. She played a huge role in teaching me how to swim. I see a lot of people now who can't swim, and I feel really grateful to her. For me, it's not just a sport, it's a means for me to feed a lot of families through diving for kai moana. I was quite a good swimmer and I made a few rep teams for it.

At school, my only competition in the pool was Blair. I was a clean-up until I went to a regional schools' competition. I was fresh off the blocks, straight into it and I was coming second, then this kid Bryden Nicholas just smoked me. He ended up representing the Cook Islands at the Olympics in the canoe slalom, which was pretty sweet. I'd never really experienced defeat like that before. Playing team sports meant that if we lost, it was as a team. When Bryden beat me in the pool that day, I had no one to blame but myself. I was gutted.

4.

A PRETTY SWEET JOB

When I was 11, Mum had started getting pretty sick again. Her diabetes had got worse. Because of her strokes she couldn't speak properly and the muscles in the side of her face were all gone. She couldn't walk at all, and she had to use a walking frame for a while. She had to take heaps of medication and some of that caused complications.

Once she realised what was happening, she started to sort out her diet and she started looking at her health. She got some rehab to try to get a bit of feeling back in her left side and to see if she could get a bit more of her speech back. Eventually, she decided to slow up on some of the meds and she started coming right a bit. She didn't stop drinking completely, but she would only drink on big occasions. Eventually, all her hard work started to pay off and she even managed to get a job in one of the kiwifruit packhouses for a while. Gypsy was working as a pruner in the kiwifruit orchards at the time as well, so I didn't worry about them as much for a while.

Blair and me stayed at Whakamārama School until the end of year 8, but Patrick got sent to Tauranga Intermediate School at the end of year 6. We didn't see him as much as we used to because he wouldn't get off the school bus until about 4 pm. By that time, Blair and me would already be up the bush making huts or down at the river.

For our last two years at the school, our teacher was Miss Holloway, who was also the principal. In that class, there were a couple of mischief girls, and me and Blair caused her a few headaches too. Her way of dealing with us was to try to teach us the meaning of 'value'. To do this, she started a points system. If you got more than 90 per cent in a maths quiz, you'd get 90 points. If you went and grabbed the firewood for the fire, or you lit the fire right, you'd get points. If you were being a dickhead, or you didn't do your homework, you'd lose points.

At the end of the term, she'd go and spend about $500 to buy all this stuff for the class. There'd be blocks of chocolate, toys, books, pens, calculators . . . there'd be two or three tables covered in all the things she'd bought. Instead of just giving them to us, though, she'd auction them and we could bid using the points we'd earned throughout the term.

She used that system each term for a year, and we all learned so much from it. We learned the importance of responsibility, punctuality and being consistent with our performance. It wasn't like she said, 'If you do this one thing, I'll give you a chocolate bar.' We had to perform well consistently throughout the term.

A lot of kids play games on their phones and they get rewards really quickly. They'll beat the first level in about 20 minutes, and that'll give them the rush of winning before they go up to the next level. Their brains get filled with all the natural, happy hormones, like serotonin, oxytocin and dopamine, so they keep on playing to get another hit. Back then, we weren't playing those games. We knew that we had to be good to earn points, and if we were being dickheads we'd lose them. We had to wait for months before we got our reward in the form of the auction. That was a really good lesson to learn.

The girls used to clean up on the points front, so they always got first choice of the prizes. It wasn't too bad for the boys, though, because the girls would buy all the hairbrushes and perfume and we still got a good shot at the basketballs and rugby gear!

It was a wonder I ever managed to keep any of my points as me and Blair were always breaking the rules or doing something stupid. It was never done in anger or bad heart or anything like that — we were just a bit high-spirited. We used to do some crazy-arse stuff together. We used to ride our motorbikes to primary school, and we'd end up getting told off for starting them up at lunchtime. Our school backed on to my uncle's farm, so we'd take the motorbikes into the paddocks and ride them. Then we'd sneak back to my house and have lunch there.

One time, we were all sitting there eating condensed milk out of a can while my auntie and uncle were at work. Then the phone rang, and old dumb-arse Blair went and answered it. Who was it? Miss Holloway. 'Hello, Blair. What are you doing answering the phone?'

Miss Holloway gave me the biggest growlings I've ever had. I've never cried so much in one year ever in my life! Whenever we did something dumb, she'd call us into her room and have a really big sit-down and talk to us. Me and Blair would sit there trying to be tough, but she'd be that disappointed in us that we'd end up bawling our eyes out. We hated knowing that we'd let her down. We'd cry so much that we'd return to the classroom with bloodshot eyes and red faces. All the girls would laugh at us for getting a growling.

Miss Holloway taught me so much. She pushed me for all my sporting events and was really proud that I was representing Whakamārama in rep teams. But she didn't put up with any crap. When I was playing basketball one time, I was going for a tip-off against a really short kid. I decided to be cheeky and, when he went for the ball, I ducked down.

She sent me off for the rest of the game for bad sportsmanship. I didn't even know what that was, so I hated her for it right through the weekend. On Monday, she explained to me what it meant. That was a really big lesson for me on how to treat other people.

For a small school, Whakamārama had amazing teachers. Lyn Harrison and Sharon Holloway played huge roles in my life. They were awesome female role models for me, and they both helped me realise that sometimes I needed to pull my head in!

The teachers at the school weren't the only ones to have a big influence on my life. One day, when I was about 11 and a half, we were sitting on the grass having lunch and this guy pulled up in a green Holden Commodore. This beautiful lady jumped out. She had really nice clothes. Then the fella jumped out and he was wearing overalls and gumboots. They looked crack-up together. Then I saw they had some sweet motorbikes on the back of a trailer. I thought, 'Who's this fulla?'

The man got some standards out of the boot of the car, and he and the lady came up to where we were sitting. I said, 'Oh, gidday, mate. What do you do for a job?'

He said, 'I'm a dairy farmer.'

'Oh, cool,' I replied. I was even more intrigued. 'What are you here for?'

'We've come to set up the cross-country course.'

I ran inside and asked Miss Holloway if I could help set up the cross-country course. After all, it was going across our farm, and I knew the route it would take because it was the same every year. She was fine with it, so I went out and offered to open the gates for the man.

He was pretty happy to have some help, and he told me his name was Ian Jeffrey and his wife's name was Lisa. Their two kids, Nicole and Andrew, went to our school, but they were a lot younger than me.

I jumped on the back of the quad bike and we took off across to the farm. As we were driving around, putting the standards in to mark the course, I asked him heaps of questions about what he did. He told me all about dairy farming and I thought it sounded like a pretty sweet sort of a job.

I asked Ian if he had any work going. He said, 'Yep! What do you want to do? We've got a piggery and a dairy farm. Take your pick.'

I said I'd work on the farm, while one of the other boys from school, Miles, got a job in the piggery.

Ian and Lisa had a big garden and they prided themselves on their roses. I helped out by doing the lawns and that led to hosing out and helping Ian down in the cow shed. From there, I moved up to helping Ian get the cows in. That was my introduction to the world of dairy farming.

GIVE KIDS A CHANCE

Looking back, I feel really privileged that the Jeffreys all put so much time into me and gave me so many opportunities. Yes, I smoked dope when I was seven years old. It made me cough so I didn't get into it. I even sold it at school. I've done some stupid shit in my life. That's not the person I am now or the person I was then. It was the situation that I was in.

There are kids like me everywhere and I'd love to see farmers give those kids a chance. You'll never know how you might change someone's life. If you see a kid in your area making an effort to get a start, talk to them. Those kids who put their head on the line to try to get work, to try to get experience, are awesome. It takes so much pride for them to make that effort, and to see them get chopped off is gutting.

All of Ian's family lived around their farm. His dad Robbie and mum Muriel lived in the flashest house in Whakamārama. They had a huge role to play in what I thought success looked like. Their place had beautiful lawns and a flash garden that overlooked the whole of Tauranga Harbour.

Robbie used to give me a $50 note to mow his lawns on the flashest ride-on lawn mower I'd ever seen and trim all the edges. For a kid, that was awesome. Not only was it good money, but it helped instil in me the mindset that farming could give you a really good life.

Muriel always used to put on a beautiful lunch, so I'd stay and have a kai with them. I think they appreciated that I could hold a conversation with them and also that I wanted to work. Over lunch, I'd ask them heaps of questions. I wanted to know how Robbie got the farm and what he'd done to get the place working so well. He'd always answer my questions and take the time to explain anything I wanted to know. He told me that you had to work hard, work with the right people and get yourself into a position to take the right opportunities when they came along. I really appreciated that.

Robbie and Muriel had the two farms, a piggery, a kiwifruit company and a grain and mulch business as well. Between them, their kids ran all of these operations. One son, Craig, ran the piggery, then Ian ran the dairy farms and the grain business, another son, Grant, ran the kiwifruit pulping factory and their daughter Gwenda did all the accounts.

The kiwifruit pulping company was really interesting. The fruit was puréed and used to produce food and drinks. The seeds and skins were used to make oil that got exported to China and was used to make cosmetics and pharmaceuticals, then all the offcut from the kiwifruit would come back to the dairy farm as feed. Nothing got wasted.

We'd also get all the pig shit and scraps from the piggery and cow shit from the dairy farm and that would go into the mulch. All of their businesses combined to help each other out. It was really smart. Robbie always used to say that being a dairy farmer meant you were in the food business and that his heart lay in producing food.

When I would go to work there for the day, I'd start work early, then we'd have morning tea at 11 o'clock, then I'd have lunch at one o'clock. Each one of those breaks was half an hour or an hour.

When I was at Ian and Lisa's, I'd see them taking business calls while I was there. Ian would be talking about the milk price, the accounts, what needed to be done on the farm. I also noticed that he always watched the news on TV every morning.

During the school holidays from the age of 11, I'd bike down to Ian and Lisa's in the dark and I'd pull into their place at 6.04 every morning. Ian always gave me grief about being four minutes late! Ian would have our cup of tea ready, then we'd go and milk the cows. We'd race to get that finished by 7.30, so we could head back over to the house and watch Paul Henry and Pippa on the TV news. Ian used to love Pippa! Even at that hour of the morning, Ian would be taking work calls.

While I spent my holidays working for the Jeffreys, my mate Patrick was working for a local farmer called Bruce Purchase. Bruce and his wife Mary had a big sheep and beef farm in the Kaimai out the back of Whakamārama.

Even though me and Blair didn't see Patrick much during the week, we always used to hang out together on the weekends. The three of us used to go up and help Bruce chip weeds, and he told us about this old possum-shooter's hut up in the bush. We went up and had a look at it, and it ended up being a deserted sleep-out in the middle of the bush. We thought it was pretty perfect, so we made a deal with Bruce. Our job was to chip the weeds in every paddock on the way to the hut, and if we did that — and our parents were okay with us going — then Bruce was quite happy for us to use the hut.

Every weekend, us boys would go up there on Friday and Saturday nights. We'd ride our motorbikes the five kilometres up the road to the bush, and then up over Bruce and Mary's driveway and across the paddocks to the hut.

In the first year, we did heaps of work on the place. We carpeted it, built a deck, extended the roof, and even dragged a set of drawers up there.

There was a little creek down the hill behind the hut, so we'd head down there to go eeling, and we'd catch freshwater crayfish. We also used to take our slug guns and slingshots up and shoot rabbits and possums. That added to our regular diet of tinned baked beans cooked over the fire!

One night, we found out that if you leave a baked bean can, unopened, in a fire, it explodes. We were all asleep and we heard this big BANG! We ran outside and there were baked beans and ashes everywhere. It must have looked amazing when it went off.

We learned a lot of responsibility from being up at the hut. On fine days, we used little tomahawks to chop down broken trees, then we'd chop up the wood and dry it in a storage shelter we'd built.

We'd leave food up there for our next trip and all our wood was there ready to go, so if it was pissing down with rain we had what we needed. We were never bored when we were up there — we could always find something to do.

We were busy kids. We never sat at home and watched TV. I was always playing sports, swimming, pig hunting with Uncle, going up to the bush. At home, I still had the pig dogs, chickens, pigs and cows to look after. There was never time to be bored. It was pretty flat out, but it was awesome.

As if I wasn't busy enough, I got my first real job. The five bucks I was earning every week was going nowhere, so — at the age of 12 — I started work at the local fish 'n' chip shop, which had recently been taken over by the guy who owned Bobby's Fresh Fish Market in town. He'd turned it into quite a flash place.

I used to help Bobby fillet the fish and I'd clean all the fish out of the bins and do all the dishes. Getting a $70 pay packet a fortnight was mean. I'd never had $20 notes before. All I thought was 'I can finally go and buy me some shoes!'

Auntie used to get all my clothes from the op shop, so having money to buy my own brand-new stuff was awesome. My mates all had mean clothes and I always wanted that. Sometimes when my friends came to stay, they'd accidentally leave their hoodies behind. I always used to try them on. I wanted money so I could buy flash gear like they had.

I was still playing rugby, but when Uncle asked me if I wanted to make the Tai Mitchell team again there was a bit of a problem. Working at the fish 'n' chip shop and having money meant that I'd been eating a lot of junk food, so my weight had gone up to 63 kilos. To be able to play in the tournament, I had to drop at least 10 kilos.

I told him I didn't think I'd make the team. He wasn't having it. 'If you want to make the team, we'll make it,' he said. 'You're going to have to go on a diet, get fit and you're going to have to start running. You've got six weeks.'

That was it. I decided I was going to make the team. I started eating healthy. I wasn't allowed chips in my lunches anymore, I wasn't allowed any bread, and I cut all the lolly drinks out of my lunch orders. At the back of our house was a really steep road, so Uncle got me running up there three or four times a day. Sometimes he would run with me, but other times he'd just follow me up in the truck.

Over that six weeks, I got really fit, lost nine and a half kilos and made the Tauranga West Tai Mitchell team. From there, I was pretty much expected to make the Bay of Plenty team to go to the annual Roller Mills tournament. It was an under-13s tournament where teams from all of the nine northern region rugby unions played each other. It had been going since 1924 and it was a really big deal.

I decided I really wanted to make the Bay team to go to Roller Mills even though I was playing crap rugby at the time. I went through a phase of being really scared on the field. I don't know why, but I was just really placid. I was hiding a lot. When there were rucks to be hit, I'd be hiding at the back of it. I'd push with my hands. I wasn't tackling, like smashing people or anything like that. I wouldn't run the ball up and I'd act tired.

When your parents are the coach of a rugby team, the trip home after a game isn't good. On the drive home after all of my games, Uncle would just tell me everything I'd done wrong. All my mates whose dads coached all reckoned it was the same for them as well.

The Tai Mitchell Tournament was in Kawerau. We had a mean team including three players who had been in the Bay of Plenty Roller Mills team the year before. It was pretty rare to get into the Roller Mills team in your first year of playing Tai Mitchell, but these guys had all done it. I couldn't see any reason why I wouldn't make the team in my second year.

We had a wicked Tai Mitchell Tournament, which meant I made the trials for the Roller Mills team. I thought I was a hot-shot. I got through the first trials fine and made the second trial team.

The fellas on Uncle's Colts team at Te Puna were all like 'Yeah, Tangaroa's going to make Roller Mills!' They all wished me luck and told me I'd smash it. I felt like making a Bay team at that young age would have given me as much mana as those 19- and 20-year-olds who were making the Bay Steamers, because I'd made the same level of representative rugby. I really thought I was the man.

The Roller Mills team was announced in the Te Puke clubrooms after the second trial. The selectors went through and read out each of the names of the 24 players who'd made the team. As I heard a heap of my teammates' names being read out, I just assumed that I'd be next.

When they got to that last name, and it wasn't me, I just broke down. I was sobbing uncontrollably. I'd never cried like that in my life. There I was in a room full of all the Bay rugby big dogs and I was crying like a little baby from the shock of not making the team.

Uncle picked me up, dragged me out of there and drove me home. In the truck on the way home, he said, 'It's okay, you'll learn from this. I didn't expect you to get in considering how you've been playing.'

Looking back on it now, I think I cried because I knew I'd been caught out at being a tag-along and letting all the other boys do the hard mahi. My mana went from way up there playing rep rugby to right down as low as it could be. I was so disappointed in myself. I felt like I'd let down my family, my teammates, my club, my school — the whole frickin' world. I wanted to crawl under the biggest rock and just stay there.

Uncle was right. Not making that team absolutely rocked me. Deep down, I knew I'd been caught slacking and I didn't like it. If I'd done as much as I could and really put the work in, I think I would have dealt with it better. I'd never failed like that before, and I don't think I've ever really failed so badly since. It really taught me a big lesson in the importance of working hard and pulling my head in. I learned that I had to be realistic about things and that I should never hide from work.

I was so disappointed in myself and I really felt like I'd let myself down, I'd let Uncle down, I'd let Whakamārama down and I'd let Te Puna down. It was a lot for me to deal with. I felt like a failure.

FAILURE

We go through life and just get on with it, so we don't often look back and think about what's shaped us to be who we are today. I reckon everyone should take some time out to have a sit-down in a chilled space to have a think about everything you've done in your life, especially your failures.

If you do that, I bet you'll have a realisation that your failures have literally shaped the way you are right now.

The biggest failure I've ever had in my life was not making the Roller Mills team. I was playing for a winning team and I should have been having a hissing season, but my mindset was one of being scared. I just let other people make the hits and take the tackles. I uncontrollably broke down crying when I didn't make the team.

Now, I realise that no one cared. I was just another player who didn't make that team. No one thought any less of me for not making that team. Most of them would have just gone, 'Oh, hard luck, mate, never mind,' then never thought about it again.

It was only me who thought of myself as a failure. No one else did. I hadn't let anyone down. I assumed that people thought that of me.

It was also the best thing that could have happened to me, because since then I've never chosen to take shortcuts, I've always done the hard yards and I've done my best to turn up and be the most haati guy possible. That might not have happened if I'd made that Roller Mills team.

For anyone out there who feels like they're failing at something and are down in the dumps, in the future, you're going to wonder why you felt like that and you're going to realise that no one is thinking badly of you. In five or 10 years' time, it's not going to matter, so back yourself and keep turning up.

5.

A VICIOUS RUGBY PLAYER

When it came to making the decision of where to go to school, Tauranga Boys' was always a big priority because Uncle Kelly went there and a lot of the boys I played rep rugby with were going to go there. Blair was going there as his three older brothers were there. That made it a pretty easy decision — I was never going to go anywhere else.

When I went to college, I didn't really hang around with my old mates from the Tai Mitchell team. I was always that guy who didn't make the Bay team. They didn't even know I felt like that, but the knock to my confidence meant I didn't think I had much in common with them. I felt like such a loser. I felt like I'd lost my mana. It was the same as if I'd been in a gang and I'd lost my patch.

After I missed out on the Roller Mills team, I adopted this

whole new mindset around fronting up and making sure I would never have that feeling of failure and letting people down again. I started training a lot harder and doing a lot more running. As a result of that, I made the Western Bay Under 13 team. In that team there were a lot of haati kids who were from Arataki, which was like the Bronx of Tauranga. They were quite well-developed and mature for their age. I knocked around with those kids a bit. We were just a Bay of Plenty sub-union team, but we used to play Auckland and Northland teams and smash them. That was really cool.

I turned into a vicious rugby player. There used to be some really big bully kids playing for college teams. They were big, strong units. We were playing one team that had this kid called Caleb on it. Everyone in the Bay was scared of this guy.

He'd get given the ball from a tip and run and he'd just smash over everyone. Whenever I got the ball, I'd just run straight at him. He didn't know what had hit him. Where the year before I would have let someone else do it, this year there was no way I was sitting back. We were like two bulls going head to head. That helped me regain a lot of respect on the rugby field. I got so much mana from that game. It was my first game back after I got dropped, and I smashed him time and time again.

School used to start at 8.45 and I'd be at the bus stop at 7.25. I was one of the first kids to jump on the school bus so it was a long trip to town every day. My routine was that Uncle would come in at about 6.50 and tell me to get up. I'd sit up, turn my light on and wait for him to leave. Then I'd go back to sleep. Five minutes later, he'd come back in: 'Hey, get up!' Another five minutes: 'You'd better get up, now!' Finally, at 7.10, I'd get up. I had 15 minutes to make my bed, get dressed, have breakfast and be out at the bus stop.

Often, my uncle would come home from work and check the mail. Hello! There'd be a bowl of Weet-Bix in there that Tangaroa didn't manage to finish before the bus had arrived that morning. If I'd had a particularly lazy week, he might even find two bowls in there. Rural life is all about early starts for young fellas and I wasn't any different.

Third form for me was quite an interesting one. Me and Blair went from being the kingpins at Whakamārama School, where we were the best at sports by a long shot, to going to Tauranga Boys' where we had plenty of confidence but were no longer the best in the school. Going to athletics days and only coming in the top quarter was something we just weren't used to.

Going from a little school to a big school was quite challenging. At Whakamārama, there were about 60 kids. At Tauranga Boys' there were about 2000 of us.

The number of people, the different types of teachers and the different ways they taught took a bit of getting used to. At our rural school, you had the same teacher for the whole year. At Boys' High, we had to get used to having six different teachers throughout the day. It felt like we were just a product and they were just sitting there sorting out the big fruit from the little fruit, and the kids just kept rolling through six or seven times a day for them. I didn't feel any sort of connection with the teachers like I had at Whakamārama.

I wasn't that good at school. I was pretty good at maths and graphics, but I was in the most cabbage English class. I'd always struggled with reading and spelling, but I never picked it up properly. I won a few speech competitions, though. I might have been rubbish at English, but I could tell a good story. I quickly overcame the fear of speaking in front of people, which has been helpful in later life.

While academically it might not have been the best for me, I loved that it was a really sporty school. The captain of the Black Caps, Kane Williamson, the captain of the All Blacks, Sam Cane, and America's Cup-winning helmsman and Olympic gold medal-winning sailor Peter Burling all went to Tauranga Boys' at the same time.

A few of my cousins from Welcome Bay were there at the same time as me. It was pretty cool that I got to see them there, because, from when I moved to Whakamārama, I'd had less and less to do with my whānau at Welcome Bay. I'd kind of adopted this idea that I was from Te Puna.

My cousins who were at the school all went to Aronui, the bilingual unit. They had the marae there, and it's where all the kōrero Māori was done. I hadn't hung around with them for ages, and it meant I had a chance to get to know them a bit more. Also, it was cool knowing my cousins had my back in a big school. Taipiri was in his last year when I was in my first year, so that gave me an unreal sense of security when I first started there.

It was weird being from two different places because I wasn't from either properly. My blood roots are in Ngāti Pūkenga out at Welcome Bay, but I feel like I left them and moved to Te Puna, where I became this Pirirākau kid.

When my mum moved away from Welcome Bay and my marae, I didn't really go out there at all. It wasn't that I didn't feel welcome, it's just that I was either at Whakamārama or at Mum's in Te Puke. Auntie and Uncle would only really go there during tangi.

Even though I was from Te Puna and I knew the land there so well, I didn't have a sense of it being *my* whenua, my tūrangawaewae. I didn't feel like I had close friends from down there because we hadn't hung out at the marae together since we were little kids. We had no stories together. In part that was because Uncle didn't really drink, so when he went to the marae he'd sit and have a cup of tea with the old people, then go home. That meant that I didn't hang out and cause mischief with the other marae kids while their parents were on the piss.

Everyone knew who I was at Te Puna, because my uncle has got a lot of mana there. I was still a whāngai kid who got whāngaied twice and the second time was late in my childhood. I think Uncle's family didn't really see me as his adopted kid, so I didn't see them as my cousins. It might have something to do with the fact that Uncle Kelly was adopted into his family at Te Puna as well, I don't know.

I knew my marae at Welcome Bay was *my* marae, but I didn't feel like it was my home even though that's where I whakapapa back to. I suppose that was because my mum wasn't there anymore. I didn't know where I was from or where I felt at home. I was a bit lost, to be honest.

Being whāngaied twice meant I didn't know where my real blood was from. I didn't have a nana and koro to ask about where I was from. I knew my blood father was from Te Kuiti and Kāwhia, and my mum was from Welcome Bay. My whāngai mum's family (and Auntie Peata's) was from Welcome Bay and Great Barrier Island, and Gypsy was from Rarotonga, and Uncle Kelly had been adopted into Te Puna. It's no wonder I was a bit confused.

At school it was the same: I was a floater kid. I didn't have a specific group that I stayed with. A lot of groups were really cliquey, and I was that one who would come in and be like, 'Chur, brothers, what are you up to today?' If they were playing cards, I'd play cards with them. I'd play rugby with the rugby boys. I'd play basketball with the basketball boys. Being from a small school, I was used to joining in with whatever was happening. I suppose that ability to adapt and interact with different people came from the way I'd always lived between different worlds and different families.

One other weird thing about going to high school was being separated from Blair. Since I'd started primary school with him, we'd spent pretty much every day together. Then when we went to Tauranga Boys' we got put in different classes. We still caught the bus together, but we both had to meet other people and make new friends. That was pretty easy for both of us as we're both really social people, but it was a bit strange for a while.

In year 9, I realised that I was a bit different from everybody else. At the time, I thought I was a bit of a weird kid because I liked doing all these things that everyone else did, but their values were totally different from what mine and Blair's were. It was all about how old our parents were and how they brought us up.

My uncle is 42 years older than me, so I was brought up in the 1970s and 1980s style. The morals that he and Auntie brought their kids up with in the seventies were the same ones they brought me up with in the 1990s. I had a totally different outlook on life and what was right and wrong compared with some of the other kids at my school. Even when I was young, I could easily go and talk to a 40- or 50-year-old and be able to have a conversation because that's who my auntie and uncle's friends were. I was able to go and shake their hands and say, 'Gidday, mate. How's your day going?' Other kids my age weren't able to do that.

Blair's the youngest of his family and his parents had the same morals and values as my family. We'd both learned to do stuff because it needed to be done. When your parents told you to do something, we knew it had to be done. I mean, we'd still complain about it and try to put off doing it, but it would get done.

From when I first started doing chores, so Uncle had time to take me to rugby, I'd had it drilled into me that when you're running a household, it's like you're running a little business. There are three things that need to happen: you need somewhere to live; you need to eat and drink; and you need to earn the money to pay for those things.

I knew that Uncle and Auntie worked really hard and I understood how the three prongs of life worked, so I would help when I could by getting washing in, doing dishes, feeding the dogs, whatever needed doing.

When I started hanging out with kids from high school, I realised that not everyone had been brought up like that. I met townies who went home and didn't do anything but play on their Game Boys and PlayStations — a lot of the rugby boys were like that.

Once I was at a mate's place in town and his mum asked him to do the dishes.

'Yeah, I'm coming . . .' he said, and then he just sat there.

'I told you to do the dishes!'

'I'm f—ing coming, Mum. Stop having a queen about it!'

I was so shocked that he talked to his mum that way that I got up and did the dishes. I couldn't believe he'd said that to her and wasn't even worried about it. I was also shocked that he didn't even get a growling for it.

That lack of consequences and lack of buy-in on his part was a real eye-opener for me.

I couldn't believe that these kids didn't see what roles everyone in their house played to make things easier on each other. Their mums would finish work, come home and cook dinner even if they were tired and just wanted to sit down. Their dads would work a hard day and come home hungry but still do stuff to keep the household ticking over.

If people only understand the eating and the drinking part of the equation and they miss out on the living and working parts, you become a needy kid. If the food and drink just arrive, you don't learn anything about independence.

If I asked Auntie and Uncle if I could go and stay at a friend's house for the night, Uncle would always say, 'Go grab some meat to take with you.'

The first time he said that I asked what for.

'They've got to eat too! If they're feeding you, they've got to pay for that food.'

After that, when I went to stay at my friends' houses for a night, I would always take a roast with me. I'm still like that. I can't go to someone's house without taking something with me. To me, someone bringing a koha like that signifies that they come with an open heart.

I used to really like staying at my mates' places in town. Being in my early teens, it was a time when girls were starting to become more interesting and boys were starting to act a bit different. I got into quite a lot of mischief with some of my townie mates, especially Matt and Poodle (whose real name is actually Steven — he got given the nickname Poodle by his dad and it just kind of stuck).

When I stayed in town during the weekends, we used to jump on BMX bikes and ride around town on night missions. That was pretty exciting for me because I wasn't used to riding anywhere where there were lights at night!

As well as me staying at their places, Matt and Poodle used to come out to the farm sometimes. They absolutely loved it! We'd go hooning around on the motorbikes and, with Blair, we'd go up to the hut and stay up there. We had so much fun just lighting fires, shooting possums and rabbits, and doing teenage boy stuff.

Some weekends, I'd go out to my mum's at Te Puke and catch up with my cousins there, but not so much, maybe every second weekend or so. During rugby season, Uncle would drop me off there after my game. I'd bike around Te Puke and hang out with my cousins at the skate park or the pools. We used to get up to a bit of mischief, while Mum and Gypsy were out on the piss.

On Monday mornings, Mum would drop me up to the bus to Tauranga. She always made sure I had a really big lunch and she'd give me spending money. She'd give me $20 and I'd just spend it that day on food. I didn't realise then that she had nothing, or I would never have taken it. She was always such a generous person, and she hated seeing other people going without.

If I was staying with Mum and we needed to go anywhere, a job for me was to go and take the warrant and rego stickers off Gypsy's car and stick them onto Mum's car. When we got home, I'd switch them back over again.

Mum's kidneys had started to pack up by now and she was getting sick again. She had to go to hospital in Tauranga once a week so she could sit on the dialysis machine for about six hours, which would clean her blood. Eventually, that turned into twice a week and then three times a week, then pretty much every day. That meant she couldn't work anymore, so I did what I could to help her when I was there. Her feet were really painful, so I'd always rub her feet while we were watching TV.

I was still playing rugby, so staying in town with my mates at the weekends meant I didn't have to get Uncle to drop me into town. I made the Tauranga Boys' top under-14s team, which meant I got to go to some unreal tournaments with them. We played Rotorua Boys', which was one of the top rugby schools in the North Island. We did heaps of training and fitness. We used to get professional coaches and All Blacks in to take skills sessions, which was awesome.

As I rocked into 14, I made the school's under-15 team and got made captain. Trying to lead the boys was a struggle for me because I was different. They were mostly city kids who spent their spare time playing PlayStation and watching TV, so I didn't really relate to them and, as a farming kid, I didn't feel like I had common chat with them. Even though I wanted to fit in, I wasn't one of the boys. I felt like I had an older mindset to them.

That year, we fundraised hard out and went down to Christchurch to play some teams. We flew down from Tauranga, which was awesome. What we didn't know was that one of the boys had taken some weed with him on the plane. He got busted smoking it in his room in Christchurch, which put me in a really weird position. As the captain, I had to talk to the team and tell them that this fella was getting sent home, but because of my background of being around it all my life, I didn't see what the big deal was.

HOW RUGBY HELPED ME SUCCEED IN BUSINESS

Playing good rugby with men and having men respect me when I was still a teenager flowed on into my work life. When I went into businesses for work, I would hold myself differently when I had to approach someone like a stock agent. I'd go up, shake their hand and say, 'How's it going, mate?' Not many 17- or 18-year-olds would know how to do that.

Growing up the way I did has helped me to read people really well but to treat them all the same. I can see a gangster who's all patched up and be like 'Chur, bro. How's it going?' and have a yarn with him. It's too easy to put people in boxes because of just one aspect of their personality instead of getting a good overall picture of who they are.

Whether you fail academically or not, you're still the head honcho at school if you're a good rugby player. Let's be honest, life is about finding out where you are among your peers and finding your space. In rugby, there's this whole pecking order that goes on, and I got a real understanding of that when I was really young. I think that's held me in good stead in business. After all, businesses are all about hierarchies and pecking orders.

One other thing I learned from rugby was the importance of how you present yourself and how you carry yourself. If you look scared, people will know you're scared. If you look like the top dog, that's how people will treat you. It doesn't matter if it's true or not.

It also taught me a lot about dealing with people. It's all about understanding the person you're talking to before they even speak. You can tell a lot about a person from the way they hold themselves, what they're wearing, how animated they are in their speech. All of those things combine to help you read people before they even open their mouths. Learning to read people is really key when it comes to doing business and employing people. Once you understand that, you can use it to make people think differently about who you are.

Everything I've learned I can relate back to sports. Sometimes, on social media, if I'm talking about my cows, I'll relate it back to an athlete. The top dog in a rugby team is always going to be at the front of a barbecue line — it's the same with cows. The bossiest one is always the first to the shed.

Putting it into that context helps people to understand concepts that might not be that familiar for them. When things are frustrating and confusing to learn, breaking things down and relating them to sport can really help.

_____

As well as school, sport and my social life, I was still working down at the fish 'n' chip shop for a couple of hours after school each day, so I used to hop off the school bus at the bottom of the road and go to work. I learned a lot about communicating with people from doing that mahi. I had to talk to lots of different people, take phone orders and make sure everyone was happy.

When I was in year 10, some of the kids on the school bus started giving me grief about stinking like fish. They'd call me Stinky Fish Hands and stuff like that. I found that pretty hard. I was also finding it hard to balance all of my schoolwork, my jobs and sport.

When I finished my shift at the fish 'n' chip shop, it was usually about 7.30 at night. I'd have a big feed of hot dogs and chips, then I'd chuck my heavy schoolbag on my back and ride my bike the 5.2 kilometres up the hill back home. When I got there, I still had to do my homework. I hated riding up that bloody hill.

One night, I complained to Uncle that I was thinking about chucking my job in, but I didn't want to do it without having something else lined up. He suggested I go down and talk to Ian Jeffrey about getting a job on the farm with him. I'd been working on and off for Ian for the past three years, so he was happy to take me on for the weekends. Goodbye, Stinky Fish Hands!

The family had two dairy farms — a 180-cow farm and a 110-cow farm. They used to employ a manager called Dion Steiner, who was a top motocross rider.

Dion's son Tyler was a really good young motocross rider, too, while Ian and Lisa's daughter and son, Nicole and Andrew, were into it as well. The Jeffreys were really community-minded, so Ian decided to let Dion build a huge motocross track in one of the top paddocks. I helped out on the tractor and the Bobcat with building some of the big jumps up there. They used to hold big motocross rides there. There'd be 40 or 50 kids all zooming up around the hill and down over all the bumps. It was so awesome.

Dion taught me a lot about the mathematics behind farming. I asked him heaps of questions about how he ended up doing what he was doing, too. He told me that he'd studied with AgITO, which was a company that had courses at Bay of Plenty Polytech in town. At that stage, I still didn't know what I wanted to do when I left school, but I really liked farming.

I started helping out with milking and getting the cows in. On my time off, we'd go over to the workshop and Ian showed me how to weld. I started working at the piggery every now and then as well.

Dion said to Ian, 'I think Tangaroa would be good enough to look after the farm during my weekends.'

Ian said, 'Do you reckon? He's only fourteen!'

Dion was like, 'Yeah! He's pretty good. At least you know he's going to turn up.'

That was all it took for Ian to give me a go.

I suppose I was really mature for my age, but I reckon that spoke volumes about who they were as people. They gave me a shot at running the little farm for the weekend. Dion wrote up instructions of how to put the wash through the cowshed and lock the cows away. He gave me a good rundown of what to do and how everything works. I asked so many questions, but I wanted to make sure I got it right.

Talk about getting chucked in at the deep end. I got up and rode my pushbike down to Ian's at 5.20 in the morning, then I'd jump on the quad bike that was at Dion's place and grab his cattle dogs. From there, I'd ride through the paddocks up to the shed, where I'd turn the lights on. I'd check on the map for where the cows were, then head down to get them. I'd use Dion's dogs to get the cows out and bring them up to the shed, where I'd crack into it and start milking. I used to get a really big buzz off it. The first time I did it, I was nervous and hoping I didn't stuff anything up. But, at the same time, I'd been working on this farm for three years so I sort of knew how it all worked.

I used to go out and pull a lot of weeds, pull the ragwort, spray the gorse, do the fencing and that sort of stuff as well. That was my introduction to relief milking, I guess!

I'd get $35 a milking, so I'd earn $140 for the weekend — and instead of stinking of fish, I now stunk of cow shit!

Even when I was working at the farm, I was still able to play rugby because I was milking first thing in the morning, then later in the afternoon.

Ian and Lisa would go away to Ohakune during the school holidays and Dion and me would run both farms when they were gone. I'd live in their house and feed the dogs.

There was a Bobcat in the grain shed where Ian would mix all the palm kernel, tapioca and barley for feed. The shed was always closed, and I wasn't allowed to drive the Bobcat because I'd never driven it before.

When Ian came back from his two-week holiday in Ohakune, he told me he was going over to mix up a load of grain for the cow shed. I said, 'Ohhh, I'll do it!'

He said, 'You don't know how to drive the thing—'

'I've already done a load. I've already mixed it!'

While Ian and Lisa were away, I'd decided to open the grain shed up and have a go on the old Bobcat. I not only learned to drive it, I also worked out how to do skids and flips on it and do wheelies.

If I wasn't working for Ian and Lisa at the dairy farm, I'd be working for Craig at the piggery. They had a contract with a couple of confectionery and baking factories where we'd get all their floor scrapings and waste lollies. We had all this palm kernel with lollies and biscuits in it that would get fed to the pigs and the cows. I'd always have a pocket of lollies that I'd pulled out as we were mixing up the loads. They were full of grain, but I didn't care. I loved eating them and never really gave too much thought to where they'd come from!

I learned how to mix that feed when I was about 14. It gave me a good idea of how to use maths in daily life — I had to know ratios so I knew how many buckets of each of the different elements to mix in to get the feed right. It was good to have a practical application for the stuff I was learning at school.

At the same time, my mate Patrick was doing a bit of work for a fencing contractor, Jason Hill, who leased a tractor shed off Ian. I crossed paths with Jason at the farm one day, and he said to hit him up if I wanted any more work.

After that, if I ever had spare time, I'd go and do some work for him as well. He used to pay $10 an hour cash, and I loved working for him because I got to work with Patrick and, at the end of the day, I'd get $80.

Another guy from our primary school, Andrew Plummer, also worked for Jason, so the three of us spent quite a bit of time working together. We used to load up a truck with a whole heap of conventional hay bales and deliver them all around the Bay of Plenty. We used to have to stack them in these weird little hay barns. I had the worst hayfever. Some days I'd go home and couldn't even open my eyes — but I got good cash for it.

I'd finish school every day and then go straight over to New World. I'd buy a big Fanta, a pizza bread, a block of chocolate and some Lynx to cover the smell of cow shit! That's what I spent my money on — well, that and clothes!

6.

A MISCHIEF KID

Poodle got his driver's licence first because he was the oldest out of all of us. He got his learner licence when he turned 15. He got a piece-of-shit car that he used to go around and do skids in.

Uncle Kelly had this old ute that failed its warrant of fitness because it was full of rust. When he parked it up and got a new ute, I asked him if I could use it to drive to work in the mornings. I was 14, driving down the road in this two-wheel-drive ute with no warrant, no rego and no licence. Going down the road, I could get it up to 80 k an hour, but driving up the hill from the farm, foot-flat top speed was 65 k. It was so slow.

I'd be driving along and hit a little bump, and I'd be able to see the gravel through the bottom of the floor. It was probably a good thing the car was so slow because I knew that if I fell through the floor my likelihood of surviving the impact was pretty high!

All of a sudden, I wasn't riding my bike anymore. I used to take that ute everywhere. I'd take it down to work for milking then drive it home. I'd pick up the boys and we'd drive up to the hut. We did heaps of paddock bashing as well. Me and the boys bought a couple of Holden Barinas for a few hundred bucks. On the farm, there was a paddock down by the river that was full of gorse. I asked Uncle if I could go and do some skids in the paddock. He said I could if I cleared all the gorse away. I don't know who got the best side of that deal!

My mates from Tauranga Boys' and the boys from up Whakamārama would turn up and soon there were a heap of cars down there going round and round, doing skids and drifting. All of a sudden, there were no tyres on them anymore because we'd popped them. We'd tow big logs around trying to clear all the gorse. We were driving right up these big hills trying to roll the cars. We did everything we could to try to blow those cars up, but they just didn't die. They were quite incredible. We made a big mess in the paddock — but we cleared the gorse like we'd said we would.

After that, I decided I wanted my own car, so I bought an automatic rear-wheel-drive Nissan Cefiro. Still no licence, but I'd been driving on the road in the old ute for about six months. I was 14 and I'd bought my first car for $1200.

I wasn't allowed, but I'd go down to the bus stop and wait for Uncle and Auntie to go to work, then I'd sneak back to the house, jump in my car and drive to school. On the way, I'd pick up some of the boys because I'd be ahead of the slow, old school bus.

On the weekend, I'd go and pick up all the boys and we'd go to the drag train in Tauranga. Some of them would be drinking, but we always had a sober driver. They'd buy the cheapest Kentucky Blue vodka and there was always someone spewing their ring out. We were never a fighting bunch, but often other people would try to fight us.

On the weekends when Ian wanted me to relief milk, sometimes I'd go out on the Saturday night and I'd tell the boys I had to milk in the morning. I'd pull an all-nighter and take the boys up with me to help with morning milking. I'd be cupping cows half-asleep and absolutely knackered, but I didn't want to let Ian down, so I never missed a single milking. We always turned up on time and were reliable, but we must have looked hilarious.

One day, Blair and me were walking up the road and we could smell something.

I said, 'Can you smell that?'

He was like, 'Yeah . . . where's it coming from?'

I jumped up and stood on top of this post. I looked over and I could see all these dope plants.

'Let's go and have a look!' I said, then we jumped the fence to check it out. It was on Uncle's farm, but I knew it definitely didn't belong to him.

We didn't smoke dope and we'd definitely never seen a plantation before. There were so many plants there but we had no idea what to do with them.

We went up the road and told one of the older guys we used to muck around with. He said, 'You guys go down and grab it all and bring it up to my house. I'll dry it all for you and we'll go halves. Then you can sell it.'

Me and Blair didn't know what the heck we were doing, but we agreed, then sussed out a plan.

We went home and told our families that we were going to be spending the night at each other's houses. Once it got dark, I jumped in the old ute and went and picked up Blair. I turned the lights off on the ute and we drove up the road to get to where the weed was. Up there, we jumped over the fence, ran through the paddock and ripped all of these big dope plants out.

With two on each shoulder, me and Blair ran back to the truck. Then Blair shouted, 'Bro, there's someone coming!'

I just about shat myself.

Blair kept dropping plants and I kept falling over as we high-tailed it out of there. When we got back to the truck, we chucked the plants on the back then Blair goes, 'What shall we do so they don't all fall off?'

'Just jump on them, bro. Just jump on them.'

We had this big pile of dope on the back of my uncle's old ute, which only went 65 kilometres an hour uphill, and Blair was lying flat out over the top of it shouting 'Go! Go! Go!'

I jumped in the driver's seat and took off with the lights still switched off because we didn't want anyone to see us.

When we turned onto the main Whaka road, I turned the lights on. Before long, we passed this car and Blair started banging on the roof.

I stuck my head out and shouted, 'What's wrong?'

'THAT WAS MUM!' he shouted at me from the tray of the ute.

I looked in the rear-vision mirror and could see that she had slammed her brakes on, thinking, 'There's the boys. What are they doing?' as she knew that I shouldn't have been driving.

I high-tailed it (at 65 kilometres an hour), foot to the floor, while Blair was lying on this massive pile of dope hoping like hell his mother didn't catch up to us.

I turned the lights off and boosted up the road into our mate's driveway. I beeped the horn as we got in there. Blair was yelling, 'Open the garage! Open the garage!'

Our mate heard us and opened the garage door, so I could drive straight in there. Me and Blair both jumped out and grabbed all the dope and ran it over to our mate's shed.

Just as I slammed the door of the shed, I looked around and Blair's mum pulled in to the driveway.

'Hi, Wendy! How are you?' I asked, trying hard not to look guilty.

'I rang Kelly and he said you were at my place. Blair, you told me you were at Tangaroa's place. Where have you guys been?'

Blair said, 'Oh, we've just been down the river, Mum.'

'You guys both need to get home!'

'Sorry, Wendy.'

After Wendy left, I turned the light on in the garage and there were dope leaves everywhere. We didn't even smoke the stuff and we had no idea what to do with it.

Our mate did a terrible job of drying it. He gave us two bags of bud and the rest that still needed drying properly. We had no idea how to do it, so we decided to stick it all between newspaper and put it in a 200-litre drum to dry. We put the drum in a big hole and left it there for a couple of weeks.

The weekend after it happened, the Edge Fest was on in Hamilton. We had all this weed that we'd never smoked, but we said we'd take it with us. We drove over to Hamilton with the air con on full with dope all along the vents on the dashboard trying to dry it all out. It was still wet when we got there, so we locked one of the cars and left it on. After the gig, it was all dry. We decided to do a hot box like we'd seen Snoop Dogg do on TV.

Not being from Hamilton, we pulled into a carpark, wound all the windows up and smoked out the whole car. Next thing, one of the boys saw some police lights. 'Wind the windows down! Wind the windows down!'

A cop car pulled out of the driveway next to us and we were sure he was coming for us, but he zoomed past and left. When we woke up the next morning, we realised we'd parked outside the Hamilton police station!

Three weeks later, we went down to get our 200-litre drum of weed, which we thought would be dry. We were really excited, but when we opened it up it was all full of rank mould. We were gutted, but we didn't really know what it was worth, so we just tipped it out and took the drum back up to the piggery. I never did find out whose dope patch that was.

That was the first and last time I dabbled in dope. I realised that I really don't like it very much.

A couple of weekends after that, we went out on the town. We had a full carload of five of us and we headed for Sulphur Point.

Out there, we used to pull into these cul-de-sacs in the industrial area by the port and do skids and stuff. There would be hundreds of cars there. We'd rip up the place doing skids. All the cars would be in a circle and everyone would watch each other.

The police would organise for a train to stop on the tracks at the opening to Sulphur Point, so no one could leave, then they'd set up a checkpoint on the other side of the train. Once the checkpoint was set up, they'd get the train to move and send a police car into where we were doing skids. The car would drive in really slowly and we'd all try to take off only to run straight into the checkpoint.

We always had a sober driver, so I never got done.

One night, I was absolutely wasted but I wanted to do some skids. I said to my mates, including the sober driver, to get out of the car while I did a few tricks. They got out, but then as soon as I started it up, they hopped back in again. I didn't move from the spot, but there was a heap of people around watching us.

The next thing some blue and red lights came on right in front of me and a siren started up. I was all smoked out so the cop couldn't see anything.

I decided I'd keep going, so the others had a chance to get out of the car. Once they were out, I stopped and the cop came over.

'Gidday, mate.'

'Oh, sorry, man, you caught me.' (I was always very respectful of the cops as I knew they were doing their job.)

'Yeah, we did. Did you have passengers in the car?'

'Nah, no passengers. Just me.'

'Why are your three passenger doors open then?'

The guys had jumped out and run but didn't bother to shut the doors . . .

'Oh, we just left them open from before.'

I don't think he believed me, but he let it slide. He breath-tested me, and I got done for having no licence, drunk driving and loss of traction.

The cop ended up asking me to accompany him to the police station. I'd never been to a police station in my life.

My car got impounded and I had four passengers there with no way to get home. I felt really bad as I saw them all standing there in the cold as I got taken away in a cop car. I fully knew I shouldn't have been doing skids.

At the police station, I sobered up instantly when they said, 'Do you want to ring your old man?'

'F—, no!'

It was about three o'clock in the morning and, as I dialled their number, I really hoped Auntie would answer the phone. I got my wish when I heard her sleepy voice. I told her what happened, and her first words were, 'I'll get Kelly.'

'No, no, don't bother him!' I said even though I knew it wouldn't do any good.

'No, you can tell him,' she responded firmly.

Oh shit.

Uncle's first words were, 'Are you okay?'

'Yep.'

'What have you done?'

I told him exactly what I'd done, and his response was, 'Well, that was bloody dumb, wasn't it? I'll come and get you now.'

He came down and picked me up but didn't growl me on the way home.

The next morning, he asked me to go and have breakfast with him. I was nervous for the telling-off that I thought was coming.

Uncle said, 'I don't mind you making stuff-ups once. No one was injured. But don't *ever* do that again.'

The police told me that I would have to go to youth court to have my case heard. I didn't know what that would involve so I rang the sergeant who was looking after the case. I told him I was really sorry for what I'd done, and he agreed to organise a family meeting.

At the meeting were Auntie and Uncle, my mum, the sergeant and Ian. It was awesome for me to be in a position where I had so much support. I hoped that by taking all of them, I could show them that I was a good kid and I wouldn't get too badly punished. We had a big counselling session about my situation. It was really good. I told them how remorseful I was, and we came up with a plan that I should do some community work.

My uncle suggested that I lose my car for three months after I got it back from being impounded for a month, that I wouldn't be allowed to get my licence for another three months even though I was going to be 15 in a few weeks, and that I go to speak in schools about my situation.

And that's what happened — I went into schools and told them about my stuff-up and how embarrassing it was. As part of it I went to a whole lot of primary schools including my old school. Having a 14-year-old come in and tell them what not to do was a bloody good idea, as it would have been way more relatable for those kids than having an adult come in.

If I'd gone to do PD (periodic detention) I would have been there with a whole lot of people who had done bad stuff and who didn't want to be there. I wouldn't have learned anything as I'd been working hard my whole life, so that wouldn't have been much of a hardship for me.

Instead I got to go to a school to teach kids about what I was doing wrong. Even though it was meant to be a punishment, I really enjoyed talking to the kids about what I'd done. It was a good way of turning a negative into a positive. I probably learned more from it than they did! Doing those talks made me realise I needed to be more responsible with my actions. That was the first and last time I ever drove drunk. While I never got behind the wheel drunk again, the whole experience didn't put me off driving. I was going to make a go-kart.

I decided I need a project while my car was impounded.

Because I'd been around motorbikes, trucks and tractors on the farm since I was really young, I had a good understanding of how engines work. I was a real tutū kid — always ripping stuff apart to find out how things worked then not being able to put them back together. When I was about 11, Uncle got back from work and was like, 'Where's the bloody microwave?'

'It blew up.'

'Oh yeah? That's a bit weird.'

What had really happened was that I'd wanted to know how it worked so I took it out to the workshop and pulled it apart. But I had no idea how to put it back together again. That happened to a lot of things — lawn mowers, motorbikes, toasters — until eventually I sussed out how they worked and could rebuild them. We didn't have the internet at home so I couldn't just google that stuff — I had to pull things apart to find out how they worked.

From the age of about 10, I knew a motor needed fuel, I knew there was a piston inside it going up and down that turned a big rotor that turned a cog and then went to a gearbox. I knew it needed oil and what the oil did. So when it came to building that go-kart at the age of 15, it cemented the idea in my head that I could build a moving item, put a motor on it, set up the fuel tank . . . the whole nine yards.

Ian had taught me how to weld and he was happy for me to use his workshop, so I took the motor out of my motorbike and used it to build a 200cc kart from scratch. This thing could go 95 kilometres an hour — it was awesome! Me and my mates used to take it out and ride it around the paddocks and I'd get the cows in on it. It was really cool.

When I got my car back, I still wasn't allowed to get my licence, so I said to Uncle that I wanted to put a manual gearbox in my car. He just rolled his eyes and said 'Whatever.'

I parked my car up in a shed out in the paddock. I jacked it up and put it on blocks, then started taking the gearbox out bolt by bolt. Again, I didn't know what I was doing, but I did know that a VL Commodore gearbox would slot straight into the Nissan Cefiro.

I dropped the gearbox out of the Nissan, and ended up buying a gearbox from a wreckers in Hamilton. I learned about how a driveshaft works, then I had to realign the driveshaft, cut it, weld it back on, and get it laser-levelled so it didn't shake. I put in a new brake line and a huge, loud exhaust on it and did heaps of work to make it look sexy from the inside.

Two months later, when it was all done, I grabbed my mattress from my bedroom, stuck it on the tractor forks, then drove the tractor over to the shed. There, I stuck the tractor forks under the car and lifted it off the blocks that it had been sitting on. The mattress was there because I didn't want to scrape the underneath of the car! With the car up in the air, I jumped off the tractor and pulled the blocks out, then hopped back on and dropped the car back down. Job done!

Even though I'd done all that work, I still didn't know if the new gearbox was going to work. I got in the car, put the key in the ignition and turned it. Bloody hell! It worked. Uncle was almost as shocked as I was.

'Bloody hell! It works . . .'

While I'd been working on the car, I'd turned 15, lasted out my stand-down and now it was time to finally go and get my licence. I got my learner's straight away, and, with the car all flashed up, I drove absolutely everywhere. I was stoked.

Meanwhile, Poodle was already planning his sixteenth birthday party. I told him I'd ask Uncle if we could have a bit of a party on the farm. He was okay with it, so we went and sussed out a location down by the river.

Once we'd worked out where we were going to have the party, we got to work on the venue. We sunk some posts into the ground and put a big tarpaulin over them to make a shelter and gathered up heaps of wood from around the farm to build a big bonfire.

My mates were pretty useless, so most of the organising was done by me. I used one of our horse floats to put the stereo in, and we even hired some Portaloos.

We told everyone at school about the party, and the whole drag train came up to it. There must have been about 300 people there from all over Tauranga.

One of the boys said to me, 'Have a swig on this!'

I don't know what it was, but it was some pretty strong alcohol — absinthe, I think — so much that the next thing I remember is waking up in my bed the next morning. I got up and went into the lounge where all the boys were sleeping. I had to ask them what had happened the night before. Apparently there'd been heaps of people hooking up, but I was more disappointed to have missed seeing drunk people falling into the holes we'd dug for that very purpose.

Once they got up, we went out and cleaned up all the bottles from along the roadsides. There was about a kilometre of bottles and boxes in every direction from the farm. We also found a bunch of Eftpos cards, a couple of cellphones and some high heels.

When Monday morning rolled around, I was lying in bed and Uncle came in.

'Tangaroa, come here!' he said.

Uh oh, he used my full name, I must be in trouble. (He never called me Tangaroa, he always called me 'My mate'.) What's happened? What's he found out?

I walked out to the living room, and he said, 'Look at this.'

On the front page of the *Bay of Plenty Times* was a photo of the party. It turned out that someone had been stabbed after I'd left. Uncle was really cool about it because he knew it hadn't been anything to do with me and there was nothing I could have done about it. It was still pretty scary, though.

When we talked to the police, we told them we'd give them any help they needed. It turned out that it had happened up the road a bit and not actually at the party, which was a bit of a relief.

One of the boys had an old van, and five or six of us jumped into it one night. We were egging cars and just being dickheads. At one intersection, we pulled up next to a guy on a pushbike. We threw some eggs at him and drove off laughing our heads off.

Next thing, there were two taxis and a police car chasing us. We turned the lights off and pulled into a church carpark. The taxis and the cop car went past, so we decided to take off again. The bro went to start his van when he remembered he didn't have a battery or a starter motor in his van.

'What are you doing, bro? Why didn't you tell us?'

'We need to crash-start it. Jump out, you fellas, and give me a push-start.'

We're in the middle of Tauranga pushing this van across the church carpark and down the road, trying to crash-start it. We finally got it going, but the cops had doubled around, and they started chasing us again. We all jumped out and ran away, leaving our mate to face the consequences.

Me and Blair got away and were walking along the side of the road. A cop car drove past, did a u-ey and stopped right in front of us.

I still had an egg in my hand, so I threw it into the bushes. The cop came over and said, 'What have you guys been up to tonight?'

'We're just walking home from town, officer,' I said.

'You guys haven't been egging any cars, have you?'

'Nah!'

'Well, can you explain that egg dripping down the side of the Pizza Hut building over there?'

'Damn. Sorry, officer.'

'You're lucky that's not illegal — but it's a dick thing to do,' he said. 'You know that guy on the pushbike you egged? That was me, mate. I was biking to work.'

I couldn't believe that we'd egged a cop! What absolute dickheads. We were so lucky we didn't get into trouble for that one.

I was still playing rugby at college and working with Ian, but I started wagging school a bit, and me and my mates were drinking quite a lot. One Thursday night, we got on the piss down at Memorial Park, then slept in our cars. When we went to school the next day, we were still wearing our uniforms from the day before. I didn't have my rugby gear for training, and I hadn't done my homework. I thought to myself, 'This ain't right at all.'

That afternoon, I went home and was heading to bed. Uncle was sitting reading a book and he called out to me. I went and sat next to him. 'What's up?' I asked.

'What's going on with you?'

'Nah, nothing. What do you mean?'

'I just feel like you're going off the rails a little bit.'

I hadn't told him what had been happening, but I wasn't hiding it from him either.

I said, 'I'm not really sure what's going on. I'm not really enjoying my rugby, and I'm enjoying going out and getting on the piss with all the boys, though.'

He said, 'Do you want to end up like all the rest of our family?'

'What do you mean?'

'If you keep doing what you're doing now you're just going to be like everyone else.'

'What's wrong with that?'

'You're just going to be average. What do you really want? Do you want to be like everybody else or do you want to be your own person?'

'I want to be successful, I suppose.'

'Well, if you want to be doing that, you need to stop doing what everyone else does and do what you want to do.'

I thought, 'Oh, whatever . . .' I didn't really understand it at the time, but now I look back and I completely get what he was trying to say.

I got influenced by how my mates were with their parents.

Often, Uncle would ask me if I could come back and help him on the farm on Sunday, and I'd say yes.

My mates would say, 'Tell him that you're busy and you can't.'

I'd say, 'I can't do that. He needs a hand.'

They'd be like, 'If you tell him you're busy, he can't make you do it.'

Once, he asked me to come and help him and I didn't go. I spent that whole afternoon thinking about the fact that my uncle was out there doing the work by himself. I knew he'd be lonely, and he would want me there. After an hour or so, it got too much for me, so I said to my mates, 'I've got to go and help the old man.'

I didn't get growled much for the bad decisions I made in life. He'd just tell me he was disappointed, and that was enough for me. It would really get me in the heart. If he'd got angry with me, I would have just got angry back. That wouldn't have achieved anything.

When I got out home that afternoon, Uncle said to me, 'You really disappoint me when you say you're too busy. When have I let you down? I go to all your rugby games. I take you to all your trainings. I go to all your school camps. If you need new school clothes, I buy them for you.

'You act like you don't want to be here. If you don't want to be here, Tangaroa, piss off. Don't be here. Go!'

I didn't want to leave him. I had actually wanted to go and help him. There was never a point where I didn't care. I didn't want to leave him hanging. He did everything for me. He was never too busy to take me places I needed to go, or to spend time with me when I needed it.

He was my mentor, and I could talk to him about anything.

I was 15 and I was a mischief kid trying to find my spot in the world. My uncle had become my best friend. It's still that way now, I talk to him about everything. People don't get him sometimes. He's really straight-up and he doesn't take shit from anyone. I'm a bit the same, so we're best mates.

7.

THE BIG SEARCH

Like most 15-year-olds, what we were going to do on New Year's was a big talking point. I suggested that we head up to the Coromandel. My mates weren't convinced. 'What are we going to do up there?'

'I don't know. I've never been there. Go camping and have a holiday?'

It wasn't the most persuasive pitch ever, but no one had any better ideas, so we decided that's what we'd do. My mates were useless — they couldn't organise the toilet paper to wipe their own arses. I ended up having to organise the whole thing. I was the youngest out of all of us, but I was the one who got things done. It was just like Poodle's party all over again.

I had to figure out where we were going to stay, ring up the camping ground, book it, work out what we'd do there, tell them what the budget was and what food and drink we'd need, tell them what camping gear they needed to bring ... you name it.

I even paid for everything, then had to chase the others up to pay me back. That taught me about budgeting on a pretty simple scale.

A group of about 10 of us packed up three cars and headed away off up the Coromandel for 10 days. My car overheated on the way up because it's such a long, windy road to get up there. I'd booked for us to stay at the Department of Conservation campsite at Waikawau.

Driving across to the east coast from Colville, we didn't pass a single car. It was just a gravel road and no people or cars for ages. I really started to wonder where the hell I was taking the boys. Then we pulled into this bay, and the campsite was packed. There were people there from all over the country. The reason no one was out on the roads was because they were all well set up and didn't need to go anywhere. It's right out in the middle of nowhere, and it was awesome. We were on cloud nine.

We took up a little boat, went diving and got some crayfish.

One day, we were at the beach and we saw this guy on a big, flash launch. He had the engines going but was obviously waiting for someone. The sea conditions were a bit choppy, so he had a bit of trouble staying in one place.

Next thing, we saw this girl, who was about our age, running down the beach and into the water. She went to jump onto the boat at the same time as he put the hammer down to get the boat over a wave that was coming. She got hit by the motor and almost lost her leg. There was so much blood. It was scary.

We pulled her out of the water, grabbed our towels and put a tourniquet around her leg. By the time the rescue helicopter got there, she had fainted. It was pretty freaky. She was lucky we were there and knew what to do. It was a big wake-up call for us.

By the time I turned 16, I'd started to pull my head in a bit. I was playing some pretty good rugby again. I made the first XV and was playing behind some pretty good players. Tanerau Latimer was the starting number seven and Luke Braid was playing number eight.

My partying days weren't quite over though.

Poodle was turning 17, so there was only one thing for it. Another party. We decided to build a shed down by the river. Uncle laughed and reminded us that the last shed we'd built was a tarpaulin on top of a couple of posts.

We decided we'd prove him wrong. A guy up the road had been milling a whole lot of big trees. We asked if we could have some offcuts and he said we could help ourselves. We took the tractor up a few times and got heaps of them and brought them down to the paddock.

Blair's dad was a roofer so there was no shortage of roofing iron or nails. We used bucketloads of nails as there was no structural engineering going on with that roof. The nails were the only thing holding it together.

We built this massive shed with walls, a roof, a bar and a fully fenced VIP area out the back. It's still standing now and people use it for camping.

Because it was by the river, the ground was quite wet, so we decided we needed some carpet or something. Instead of getting carpet, we decided to drive down to a highway gravel pit at

Ōmokoroa to nick some aggregate.

The only problem was we didn't have a trailer. Then I realised there was one that had been parked up in the paddock for about 10 years. It was a heap of shit. It was missing a tyre and didn't have a towball, so we tied it onto the Cefiro using some rope. Since I didn't know how heavy gravel was, it seemed like a good idea.

We went down to the gravel pit and we loaded this thing up. There was no way we wanted to do two loads, so we probably put about a tonne of gravel in this old trailer.

As I drove home, I thought the trailer was pulling to the left a bit and sparks were flying off the rim of the left wheel where the tyre should have been, but we got back in one piece. We unloaded the trailer and put all the gravel down on the floor of the shed.

I persuaded Uncle to get us a load of concrete from his work, which he did, and we poured that over the gravel, so we had a thin concrete floor.

All we needed then was power. We decided we needed a generator. That was the one thing we trusted Poodle to organise for his own party. He said he'd go and hire one. We should have known better because Poodle was the least organised out of all of us.

The other boys and me had made this huge bonfire. We had a big stereo set up, and we had heaps of alcohol in our VIP bar. We sold VIP tickets for $15 each, and they had to show them to a bouncer to get in. There was a barbecue out the back and free booze for everyone in the VIP area.

It was still daylight when people started to turn up. There was probably about a hundred people there and we had no stereo.

Someone rang Poodle and asked him where he was with the

generator. 'Yeah, bro, I'll be there soon, I'll be there soon.'

Eventually, Poodle and Matt turned up with this big-arse generator. I pull-started it and it went to take off.

'What is THIS?' I asked.

'I don't know. It was next to the generator,' Poodle says.

'Bro, this is a compactor.'

'Oh stink, is it? Where's the generator?'

Matt goes, 'The generator's over there,' and points to what is actually a generator.

'Why'd you get a compactor, Poodle?' I asked, genuinely confused.

'It was just on the back of the truck with the generator.'

I still have no idea what the hell he meant, but at least we had power for the stereo.

We'd dug a big hole in the ground for the generator so it wouldn't be too loud. We started up this big 6 kVA generator and cranked the sounds and got the lighting going. We lit the big bonfire with a fire cracker. It went 'Whhooosh!' and was burning strong really quickly.

There were about 350 people there and the drag train turned up again. There were cars everywhere. It was massive.

Auntie and Uncle came down to check on us. Uncle couldn't believe his eyes. 'This is like a big concert, bro!'

Everyone had a really good time. There were no fights and we didn't end up in the paper.

The next day, we went down and there were still people drinking. We went up and started cleaning up the bottles along the road. We found this really flash car crashed on the side of the road. It was a bit weird, so I rang the police and gave them the registration number. It turned out that someone had stolen this brand-new car from a showroom in Auckland and driven it down to the party, where they crashed it and left.

While I was in year 12, I didn't really like college that much. I liked the school, the teachers and being with the boys. I was good in art, graphics and maths, but I could not get my head around science, English went straight over my head, and I wasn't very good at geography.

It wasn't that the school system was bad, I just didn't understand how it worked. When you don't know the value of something, how do you value it? They'd be telling us all this stuff, but I couldn't see how it applied to my life or how it would be beneficial to me to understand any of it. There was a lack of connection between what we were being taught and the reality of my life.

By the time I was 17, Mum started really going downhill. It had got to the point where the dialysis wasn't enough. That year, she started going over to the hospital in Hamilton every four days to have blood transfusions. They put a port in her arm so she could get blood transfusions done more easily. She'd come back feeling really good, then after three days she'd start to deteriorate because her blood wasn't being cleaned properly. She had to take heaps of pills and she still couldn't use half of her body. She had real bladder trouble, so she ended up having a toilet in her room.

Once I was earning a bit more money, I'd give Mum money or buy her heaps of food on the weekends when I stayed out there.

I used to say to her, 'Are you all right, Mum?'

She'd say, 'Yep, we've got kai in the fridge.' That's how she measured whether she was well-off or not.

Some of my friends didn't understand why I gave my mum

money. They were like, 'You shouldn't be giving your parents money — they should be giving you money!'

I really struggled to understand that mindset. I knew that by giving Mum money I made her life a little bit easier and that was really important to me.

The one thing I did understand was the importance of making a good income. I'd watched my mum and Gypsy struggle so much living week to week trying to make do on Mum's sickness benefit and whatever Gypsy managed to bring in. I knew I never wanted that for me or my family.

I was still working for Ian, but I'd started to think seriously about what I wanted to do with my life. It was then that I remembered Dion saying he'd studied with AgITO. I decided to go and see my careers advisor, Mr Meys. I told him I really wanted to leave school. He suggested I try to get into the college's work academy. I'd have to start off doing Gateway, then, if I was good at it, I could see about getting into the academy.

GATEWAY AND INDUSTRY TRAINING

The Gateway programme is available through a lot of schools. It's for kids in years 11, 12 and 13 who want to get some practical work experience while they're still studying for NCEA. One of the best things for me about Gateway was I got to work towards my farming qualifications while I was still at school.

The main organisation that offers courses in dairy farming is Primary ITO. They're the industry training organisation for anyone who wants to work in farming, fishing or horticulture.

Most of their courses have quite a lot of on-the-job training, and they cover just about every aspect of dairying you can think of — even goats.

The Smart Start course is a good taster for anyone who thinks dairying might be for them. It's a seven-month course, but you come out of it with your level 2 New Zealand Certificate in Primary Industry Skills in Agriculture.

From there, you can go on and do level 3 certificates in more specialised things like vehicles and machinery, milk harvesting and livestock husbandry. Most of these courses take less than a year.

After level 3, you can either go for your apprenticeship, or take on your level 4 certificate, which takes about 18 months. To do either of these, you need to be working in a job where you have a bit of responsibility and can learn a lot. After that you can get into studying your level 5 in primary industry production management or do a diploma in agribusiness.

All of these courses teach you heaps and help set you up for a career in the industry, so I really recommend getting amongst them if you think dairying might be for you.

The Gateway programme meant I still had to study for NCEA, but I also got to do work experience placements and be assessed on what I'd learned out in the real world.

As well as farming, I was interested in architecture because I was good at graphics and was also into audio electrics. My mates in town would buy really flash stereos and subs, and I'd install them and do all the electrics in their cars. I really liked the challenge of hiding all the wires and listening to make sure the sound was right.

I ended up getting into the academy and we had to do a project called 'The Big Search'. It involved choosing three industries that you wanted to work in. I had my three all lined up — audio electrics, architectural design and farming.

We had to do some research, which involved ringing up a professional in each of the industries and asking them questions, like what the starting rate was, what you'd expect to be earning in five years' time and what would it take to get there.

I did all of this research and farming came out on top. After five years as an auto electrician, you'd be in your second year of qualified trade and you'd be on, say, $22 an hour. With graphic design, you'd just be out of university and you'd owe maybe $40,000 in student loans. If you were in farming for five years, you'd have your level four New Zealand Certificate in Agriculture and your apprenticeship, you could be earning $85,000 and potentially running your own business. That sounded pretty mean to me — and I already knew what the job involved.

My tutor in the academy was Mr Sperling, and he was a big help to me. The cool thing about the academy was we had a classroom that had a pool table, a foosball table, a kitchen, an old VW van that doubled as the computer room, a couple of couches and some desks.

The structure of the academy was that you could do whatever you wanted for the time you were at school. You could eat when you wanted and you could make a coffee when you wanted, but you still had to complete a specific amount of work in a set amount of time. It was based on credits, so the harder you worked the more time you got off. You got out what you put in, which is how it is in the real world.

I really liked it because I knew that if I went hard with my work, I could have a day off or I could go and play pool for a bit.

We had a different uniform — the academy kids always wore a tie, a white dress shirt and a blazer. We were very well presented because we were going out to work placements twice a week.

My first work placement was up on Ian's farm, so I was basically getting credits for doing a job I already had. I got my level two with AgITO, which changed its name to Primary ITO in 2012, and started my level three while I was still at school.

While I was at the academy, I was still in the first XV, so I had practice on Tuesdays and Thursdays. I was going to work on Friday, Saturday and Sunday (and sometimes Thursday if I'd got my schoolwork done by then). Doing all that physical work chipping weeds, doing fencing for Jason, getting cows in and running up the hill meant I was always really fit for rugby, and I reckon my work helped with my mental toughness as well.

Once I made it to year 13, heaps of my mates started dropping out of school. I wanted to do that as well, so I decided to ask Ian if I could get a full-time job with them. Even though I'd been working for him for about six years by then, I knew taking me on full-time was a big leap for the Jeffreys.

They agreed to take me on. I was stoked. Before leaving school, I turned 17, finished the rugby season off, went to the seventh-form ball, then I got my leaving certificate and I was out of there.

Rob Sperling mentored me into the role and helped me

negotiate my contract. One of the biggest hurdles young people have when they're leaving school and transitioning into work is that you all of a sudden have to work with a bunch of adults and you're at the bottom of the pecking order. It doesn't matter who you are at school, you start at the bottom anyway (unless you're a rugby player!). It was really cool having Rob there and I knew that he was going to support me outside of school so I never felt alone. I really appreciated how hands on and helpful he was to me. For quite a while after I left school, he'd check in on me to make sure everything was working well for me. I still talk to him quite often now, so the impact he's had on my life is huge.

MENTORING

I was very fortunate to have Rob Sperling supporting and mentoring me when I left school. Having someone to push your barrow and mentor you is huge.

If you're looking for a mentor, find someone you look up to and who you think will be able to help you to get where you want to go. Once you know who that is, just go and ask them.

Tell them that you look up to them and that they have lots they can teach you, then ask how they would feel about being your mentor.

It might feel a bit scary to ask, but the worst thing that person can do is say no. Most people will be really stoked to even be asked, and will be happy to share whatever knowledge they can and help in any way they can.

If someone asks you to be their mentor, ask yourself – and

them – what value you can add for them and their career. That way, you'll both have a clear idea of what it is they need and what they're trying to achieve.

If you see a young person who is heading out into the world for the first time and you reckon you can help them, all I can say is step up and do it.

Mentors hold you accountable. When you're working towards big goals, it's easy to go off-track, so having someone working with you and mentoring you will help you stay on your path.

8.

ONCE YOU'RE UP, YOU'RE UP

When I got the job with Ian and Lisa, I moved out of home. Ian and Lisa had built a little two-bedroom cottage out the back of their place because they knew they were going to have to start employing another person for when they decided to step back a bit. I had the place to myself, but I was only three minutes up the road from home. I was pretty much living right outside the cowshed, so I got to have a little bit more sleep in the mornings, too. It was wicked!

Living on my own meant I had to do all my own cooking. I'd never seen a man cook for his family until I met Ian. My uncle used to cook a mean feed, but it would be basic stuff like a steak. He didn't care about it being pretty. When I first started working for Ian, some days he'd say, 'Do you want some scones?' And he'd just smash out some really beautiful scones. He'd make up the mix, cook them, then we'd eat them within that hour. Seeing a man bake and be really quick with it was cool for me. I'd be in there, asking him questions about what he was doing and what he was putting in them.

Whenever me and my friends went away up to the Coromandel, I was the one who sorted out what we were going to take and what we were going to eat. I used to end up doing a lot of the cooking. Up at the hut, the boys would heat up canned muck, while me and Blair would knock up huge feeds.

When I moved into the new house, Ian and Lisa bought me a slow-cooker. I had a huge fridge-freezer with all my meat in there and it was really cool. I'd get all my meat from the farm. Every so often, Ian would get a couple of sheep from down the road shot. Me and Dudley got the job of hanging them up, then we'd take them down to the Whakamārama butchers to get processed. We used to get these big rolled roasts, and mutton flaps with heaps of hinu on them.

Every now and then, first thing in the morning, Ian would get me to go to the shed and grab a roast out for dinner. I'd watch him put rosemary in the lamb and prep all the ingredients, then put it all in the slow-cooker. By lunchtime, this thing was smelling so good I couldn't believe it. I never cooked at home, so getting to help Ian was how I learned.

BABOOOOOSH TERIYAKI CHICKEN
This is Court's favourite dinner and it's super easy and tasty as. It's also real good for impressing people, so if you're

cooking for your new girlfriend or your in-laws are coming for tea and you want them to think you're the flashest fella in town, this is the way to do it.

INGREDIENTS:

Rice
2 cups medium grain rice
4 cups water
A whack of rice wine vinegar

Chicken
Boneless chicken thighs
Olive oil
Flour
Masterchef Teriyaki marinade

Salad
Tomatoes
Avocado (if you're rich or in the North Island!)
Cucumber
Lettuce

Toppings
Soy sauce
Japanese mayo
Toasted sesame seeds

Utensils
A frying pan
A rice cooker
A big platter for serving

METHOD

1. Before you start cooking, make sure you clean the cow shit off your fingernails.
2. Chuck your rice, water and vinegar in the rice cooker and set it going.
3. Chop your chicken into chunks. Smack some oil in a frying pan then heat that puppy up.
4. Dowse your chopped thighs in some flour so the sauce sticks to them.
5. Biff the chicken into the hot oil but don't fill the frying pan up too much. Flip the chicken until it's cooked through – if you don't cook it enough you might end up getting a week off work with salmonella.
6. Pour the teriyaki sauce onto the chicken then use a brush to smooth the sauce all over it until it's got a real good covering and is all nice and sticky.
7. Chop up the veges for the salad then cover about half of your platter with them. Squirt some soy sauce in a little bowl and sit it amongst the salad so you can use it as a dressing as well as dipping your chicken in it.
8. On the other half of the plate, pile up the rice then put the delicious sticky chicken on top of it. Then get that Japanese mayo out and squeeze it across in stripes until it looks like a chequerboard, then chuck on the most expensive, flashest sesame seeds you can find.
9. Annnnd bloooody smash it!

Mum was the loveliest lady you could ever meet. She was the manaaki queen.

Me, gumboots on, in the pig pen at Whakamārama.

Early farm experience. Me, Uncle, Sheba and cows.

My nana and all my first cousins: Jonathan, Ihakara, Taipari, me
and Nana Mōkai, and Jarrod.

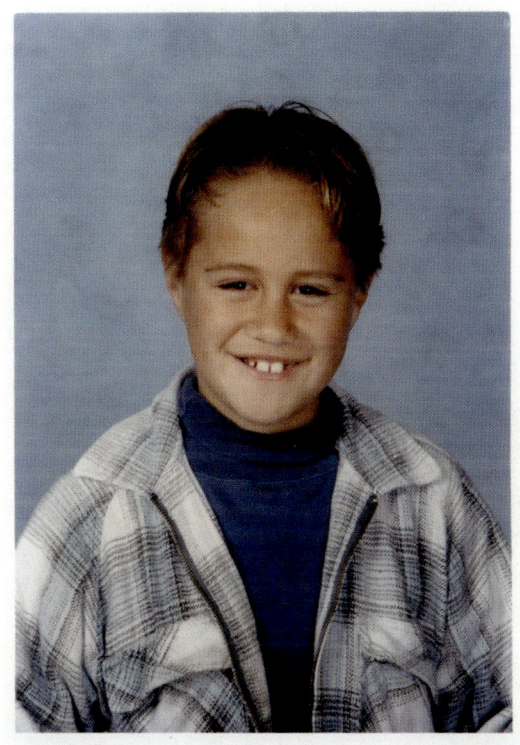

Whakamārama
School was my sixth
primary school.

My name was
changed from Dylan
Walker to Dylan-
Tangaroa Walker
because I loved the
moana so much.

Te Puna J.M.C. 2002. There was a really good culture at Te Puna, and a lot of their mana came from their rugby team.

Pig hunting with Andrew in Ōpōtiki.

My intro to the world of dairy farming started with helping Ian and Lisa with their lawns, which led to hosing out the cow shed . . .

Awards night! In 2012 I was named Primary ITO Dairy Trainee of the year.

Me and Courtney started up our gym, The Barracks, in the old Army Drill Hall in Invercargill.

You are who you surround yourself with, and if you're surrounded by dogs for eight hours a day their positivity rubs off on you.

Even when I was little I used to think about being a dad, and the things I'd do with my kids.

Uncle and Auntie were super excited to have their first grandkid, Tekauenga. Now they want more!

We have an unreal team. Here's me, Detroit, Guri and Mikayla.

Diving is my best hobby. It's not just about getting fish for kai — it's also the experiences, taking the plunge and taking time out.

I used to eat a LOT but, because I could cook, I loved eating at home. As well as my roast meals, I used to eat heaps of fish as well. Uncle and me still went fishing together a lot. We'd take the flounder nets down to Ōmokoroa, we'd get oysters from down on the mangroves, and we'd get pipi and scallops, then cook it all. It was bloody awesome.

The biggest thing I discovered was how much I loved farming and loved being outside. I couldn't imagine anything better than being out there right now. I loved being on the farm and working with Ian, so the move from being at school to working full-time wasn't hard for me. I'd been doing it for a long time, anyway. The mornings aren't fun, but — as much as they suck — once you're up, it doesn't really matter. Even now I hate getting up early, but once you're up, you're up.

I was enrolled with AgITO, so I used to drive out to Te Puke, where their classroom was, every fortnight to study.

Ian was the type of boss who would give you all day to do a job. I loved knowing what my jobs for the day were, and being able to just go out and do them. Ian let me timetable my own day and put trust in me that everything would get done.

By that time, Dion had moved on from the Jeffreys' farm and Dudley Thorburn had started working there. I worked a lot with Dudley, who was about 40, and he was a bit of a mentor for me. He used to tell me about all his life experiences. He told me about when he'd worked on a big beef ranch in Montana in the United States. One of his best yarns was one he told me about going bear hunting.

I used to think 'Oh, this fella's talking shit' because he had so many loose stories. One day, I said, 'Oh, whatever, man!' and told him I didn't believe him.

Dudley said, 'Well, come and have a look in my photo album!'

I went over to his place and had dinner with him and his wife Sarah, and the photo albums came out. It turned out he'd gone to the United States on a farm exchange and all the yarns he'd been telling me were true! When he told me about his experiences, it really opened my radar to travelling and all the other opportunities that farming could give me.

Dudley was a big part of my life when I was starting out. When you're just learning from people, you can be easily misled. If I'd been in a gang, I would have done things the way that gang did. It was the same with farming. I was in a period in my life when I was just sucking in so much knowledge and he taught me a lot about life skills, farming, travelling and getting life experiences.

Until I met Dudley, I'd never really thought about going travelling, but he really made me want to go overseas. I've always wanted to go to Montana to see where he went.

Dudley was the most chilled-out guy you could ever meet, and I think he loved having a mate to work with. He was really funny — a bit of a different character — but I used to ask him so many questions and he was always really patient with me. He taught me a lot of stuff.

One thing he taught me was how to train a dog. Because he used to work on sheep and beef farms, Dudley had really good stockman skills. He was one of the calmest people I'd ever met, but haati. He was a tall, wiry guy. He had a really good dog called Jess — a little grey thing that was awesome with the stock.

I had always wanted a dog. Because I'd been feeding and looking after Ian's dogs, Max and Jess, when he was away, I'd learned a bit about looking after dogs quite early on. When I'd been working for Ian for three or four months, I decided I was ready to get one of my own.

I had a look on Facebook and saw one that had come up for adoption from the pound. It was a Collie Lab cross. I ended up going to the pound and I had six dogs to choose from. One of them was the one I'd seen on Facebook. He was only a puppy — about five or six months old — and he was sitting down on his butt, wagging his tail.

I bent down to pat him, and he got so excited he started pissing — right into my gumboot. I didn't notice it until I felt it. The little hua!

I just thought, 'You're the one' and I ended up getting him. I named him Chase, and he became a huge part of my life. He's got me through a lot of tough times on the farm and he's still with me today.

I brought him home and I had to teach him how to do everything. I had to teach him not to bark and not to dig holes, and all that sort of stuff. It was really cool to have such a huge responsibility. Having a dog taught me a lot of discipline.

Dudley showed me how to teach my dog alongside his one. We ended up tying the two dogs together and Chase would follow Jess through everything. When Dudley would tell Jess to get away back, Chase would have to go, too. That's how he got to understand all the commands and whistles.

I ended up with a really good dog that was really well-trained. Chase is a super intelligent dog, and I'd taught him not to dig. He wouldn't do anything bad like rip up socks. He knew the difference between left and right. He knew how to go back and push up. He knew how to pull cows out and bring them to me. He was just a really good dog.

One day, I hadn't been home long, and he started digging this big hole right out the front under the deck, the little hua. I grabbed him and rubbed his nose in it and said 'NO!' He took off to his bed for a good sulk.

The next afternoon, I came home, and he was bloody doing it again!

'You little prick! How dare you do this to me?'

If he'd been a person, he would have been crying because he was one of those dogs you could just look at and he'd drop his head because he knew he was in trouble.

On my next weekend off, I decided to go and fill the hole in. I got under the deck and went to fill in all the dirt and, right by where he'd been digging, I found his favourite ball. That was when I realised that when I got home on those days I'd said to him, 'Where's your ball? Where is it? Go get it!'

Chase had then taken off and started digging. I didn't know he was looking for his ball — I just thought he was being naughty all these times. My poor mate! I felt bloody stink. I went over and apologised to him and he looked even more confused.

Ian had a run-off property about a kilometre down the road, on a farm belonging to a guy called Tommy Lockhead, opposite the Dwyers' place. They were really close family friends of ours. Peter Dwyer owned a scaffolding business — he fell off a 10-storey building and broke every bone in his body. After that, he ran a business putting up shade cloths. I really looked up to his boys, Benny and Stan, who were about two or three years older than me, because they were into the boy-racer scene. They had wicked cars that were a lot flasher than I could afford.

The support block was on one of the steepest hills in Whakamārama. You wouldn't drive up it on a quad bike or a two-wheeler. We had about 150 sheep down there as well as all our rising one- and two-year-old calves. We'd rear them there to replace cows that were getting culled.

Dudley, Ian and me used to drive down there every second or third day to shift stock or if we needed to muster them up and put them through the yards. On the hilly block, the dogs got a lot of work. It was good fun teaching the dog down there with the sheep.

I worked with Ian's brother Craig a lot as well. I used to look up to him. He was a good-looking fella and he was really hard out. He wouldn't mess around at work. He had a Honda quad bike that he always used to do skids on. He'd do snakies all the way up the paddock and he'd boost it home, so obviously I thought he was the man!

I was surrounded by really good men. I had Ian, who was a businessman; Craig, who was a really styley, hard-out fella who'd have your back for anything; I had Dudley, who was a really chilled-out fella who loved a good yarn and had had incredible life experiences; and of course I had Uncle.

When I was working for Ian, I still used to go down and help my auntie and uncle on the farm. The cool thing about working with the Jeffreys was they'd let me use some of their gear to help out with work down at home. I'd take the tractor over and mulch the farm or do whatever Uncle needed help with. Equally, if Uncle ever needed to borrow anything, Ian was happy for him to come down and grab it. That made life a lot easier for him.

When I left college, I went to play rugby for the Te Puna

premier men's team. I played at number seven as a loose forward, so there was a lot of tackling involved. I hadn't really developed muscle yet, so I was only about 83 kilos. Playing against men took a bit of getting used to as I was only 17, but my fitness meant I blew everyone out of the water.

I felt like I'd finally come into my own and was playing some pretty good football. The boys told me I should have a crack at the Bay Colts, and I was like, 'But I'm only seventeen . . .'

Instead, I trialled for the Bay Under 18s. I figured that would give me two years in the team before stepping up to try out for the Under 20 Colts team. But straight after the trial for the Under 18s, they had a trial for the Under 20s and I decided to have a go at that as well. I was pretty surprised when I made the team. I had finally made a Bays team. Even better, it was one that I really shouldn't have been in!

I ended up playing for the Bay Under 20s when I was only 17. Our coach that year was Jeremy Cotter, who went on to coach Manawatū. Cotts was the man. Me and him got on quite well because he was a farmer as well.

We used to have our training at Eastern Districts Rugby Club, just out of Te Puke. I had to drive about 50 minutes from Whakamārama after work. Ian was really supportive of my rugby; he knew my story and he knew how important it was to me, which was good. I'd leave the farm at about 4.45 pm, pick up half the boys who didn't have cars, then turn up to training.

A couple of times at training, we had Paul Tietjens in for fitness training. He used to smash us. Being pushed by someone like Titch took my body to another level. It showed me how far I could go. He would break me and it was wicked in terms of my stamina.

On the way home from one of his sessions, we had to stop in Te Puke because I thought I was going to pass out, I was so stuffed. At the time, I was getting up at 5.30 am to go to milking, then finishing work and driving all the way to Eastern Districts, then training with Paul before driving all the way back home again. I was rooted.

I was still driving the old turbed-out manual Nissan Cefiro, but I really wanted to get a truck for work. Ian had an Isuzu Rodeo, which was a really cool-looking beast of a thing with big, wide tyres, tinted windows and a big flat deck on it. When he said he was going to upgrade it for a new one, I said, 'Ohhh, are you selling your old one?'

He said, 'Yeah, I want ten grand for it.'

I thought, 'Bugger. I don't have ten grand.' I was only 17 and just out of school.

A bit later, I told Uncle about it. Straight away, he said, 'Shall we go get you a loan?'

'Oh, yeah! What does that look like?' I had no idea what was involved.

'You just go to the bank and get a loan, then you repay it.'

He then took me through the experience of applying for my first loan. Uncle went guarantor for me and I had to get some life insurance to cover the loan, but next thing I knew, I was 17 and I owed the bank $8000. I sold the Cefiro to make up the extra two grand.

The whole process was a really cool life experience. Having to get a loan, then having to pay it off every month and giving $10,000 over to Ian in cash and getting my first truck in return was wicked.

The truck was mean as, with all the bells and whistles on it. I used to take it over to rugby training, and it was a much more comfortable drive than in the Cefiro all loaded down with the boys.

Our Under 20s that year was a really good team. They were a really fit bunch of boys. When I'd made the move from club to schoolboy rugby, I went from playing with a lot of these guys in the Western Bays teams to playing against them for Tauranga Boys'. It was really good to be back playing on the same team as some of them. We used to travel all over the North Island to play against other rep teams, which was awesome.

It was about that time I got exposed to the drinking culture of rugby. I'd been drinking since I was in my teens, but this was different. Now I was going out night-clubbing with men and seeing some of the stuff they got up to.

The women aspect of rugby was really bad. It's not bad when you're the boy in it, but looking back on it, what rugby boys do and what it's all about is pretty gross. It was like a pissing contest over who could root the most sheilas. It was like a weird sort of competition that came from a strange combination of immaturity and insecurity.

All the way through rugby, you'd see men who weren't loyal to their partners. The better they were at rugby, the worse they got. It's awful for the partners. All of their responsibilities at home would just go down the gurgler. If you were a little bit niggly about it, and said it wasn't your thing, you'd be classed as weak. There was a real pack mentality about it. It built a real insecurity in some of the younger players, who would get into it really half-heartedly to avoid being picked on. Everyone was loyal to their teammates instead of their partners. It's a bad culture that doesn't get talked about enough — there's a real bro code about it.

But then there'd be the women who'd jump all over us rugby players, and they knew that we were taken. It's almost like rugby players are trophies to them, too. It's just so toxic. Those women would often end up becoming the guy's new partner, then the same thing would happen again. It was a really weird and quite gross cycle. It was a big eye-opener for me. My only experience of it before then was from being in the first XV, but that was just schoolkid stuff. A lot of these guys were married and with kids. I just didn't get it.

It was pretty real when I started to see husbands and wives splitting up and families getting torn apart. Even while I was out there in the thick of it, I found it really sad. It made me really aware that sometimes people you think you know aren't who you think they are. Sometimes they can be a sheep dressed as a wolf and vice versa.

I guess hanging out with older guys at rugby got me thinking a bit more about what I wanted to do in my life. One day, Ian and me were in the cowshed. We'd just finished calving, and I turned around and said, 'How do I get like you, Ian? How do I step up and become a farm owner.'

He said, 'You have to go where the grass is greener and it grows the cheapest.'

'Okay, where is that?'

'Probably down in Southland.'

'Where the f— is that?'

I don't know how people go farming without a dog, to be honest. My dogs motivate me to get out every morning. I'll crawl out of bed and they'll be waiting there, shivering with excitement and ready to go. They'll wait just long enough for me to get a coffee in me before they start to get a bit wound up about getting going. They really help me get into my day.

Dogs have awesome personalities. They're more high than anything, but they're never low. You look at a dog and, if anything, their low is a neutral. Where we humans can get really low and down in the dumps, a dog will always go back to level. It's really rare to see a depressed dog.

You are who you surround yourself with, and if you're surrounding yourself with dogs for eight hours a day their positivity rubs off on you. All they need from you is a bit of praise, and when you give praise it makes you feel really good. Giving praise — if it's to dogs or humans — gives you a sense of fulfilment, happiness and love.

If you don't go out of your way to give a dog affection or praise, they'll come over and ask for it. A dog will put their nose on your leg wanting a pat. If you give them a pat or a hug, that affection can make you feel really good.

They have such a good sense of how you're feeling that they know exactly what to do. My dogs know when I'm angry, and they'll make themselves scarce or they'll sit there quietly. If I'm feeling down or crying, my dogs will always come in for a cuddle, which makes me feel heaps better.

MY THREE MUSKETEERS AKA MY DOGS

Manawa is hearty, like his name. I got offered $600 for him when he was only about three or four months old. I was into it but Courts said, 'Hell, no! I'll sell you before I sell him.' As usual, she was right. He is the mantis and saves my life all the time. He's really good at getting cows off crop and whatnot.

Waha lives up to his name because he's really loud and has a mean bark on him — but trying to get him to bark at the right thing can be a bit tricky. He'll bark like crazy if someone drives up the driveway, but try to get him to get some cows in and — oh hell no! — he doesn't know what he's up to.

When I was training him, I'd take him down to the paddock and I'd be like, 'Get away back!' and old Waha would be like, 'Eh? What's that, bro?' Then he'd piss off and jump in a trough and sit in there and sulk. Waha is pretty much only good for turning good dog food into dogshit and for leading the other two astray.

Chase is an old man now. He was the meanest dog when I first got him. It was love at first sight for me while he pissed in my gumboot. We've never looked back since then. He's been my mate for the past 16 years and he's got me through a few scrapes. I've been trying to get him to retire as he's slowing down a bit and his joints are getting a bit sore, but he's not having a bar of it.

People on dairy farms who don't have dogs often don't like dogs. It can be hard for some people to understand how they can help you work with cows. If you've got a dog and you've seen what they can do when you're working with animals, you realise how awesome they are. The dog can change the way animals think about things.

The cows up on a dairy farm are used to dogs, but they still see them as a threat. When they hear the bike come, they'll all get up and come to the gate. When there are no dogs there, the cows will all just go up to the gate and stand there until we physically move them. We might have two-thirds of the herd not moving and all we can do is push from the back. All that really does is just get the back cows into a big jam with the ones at the front.

If the dog is there, the cows are aware that there is a threat, so they'll walk through the gate and onto the lane. It's way easier to get them to all move at the same time. It's not so bad during the winter and spring because they want to get to the shed to have a feed and get milked, so they'll go a bit more willingly. In summer, when it's hot, they'll start winding down. Often, they'll be heavily pregnant so they're not producing as much milk, so there's not so much of a desire in them to go up and get milked. A dog is really handy to get them moving where you want them to go.

A TOWNIE'S GUIDE TO DAIRY FARMING

Cows get milked from August to April, with the peak of the season usually happening in late October. From May to July the cows have a bit of a break from milking in what we call 'drying off'. Then they'll usually calve again starting in

August. Over the drying-off period, some farms keep milking to supply the local shops with milk.

Most farmers milk twice a day, once in the morning and once in the late afternoon, but some of them are moving towards milking only once a day. On some farms, they even have a kind of robotic milking system so the cows can just take themselves to the shed when they feel like being milked.

New Zealand produces about 3 per cent of the world's milk and is the eighth-largest producer globally.

There used to be small dairy factories in most rural towns around the country. They gradually merged into bigger and bigger co-operatives, until two co-ops dominated the country — New Zealand Co-operative Dairy Company and Kiwi Co-operative Dairies. In 2001, they merged to create Fonterra, which now processes nearly 95 per cent of the country's milk. There's a few other smaller co-ops out there, but Fonterra is by far the biggest.

The way it works is that the farmers who supply milk to the company are also its shareholders, so when things are going well, they get paid well. The payment they get is worked out by the company and is paid on the basis of a certain amount per kilogram of milk solids — that's the fat and protein that are left when all the water is removed. That price can vary quite a bit depending on quite a few things, including international demand for our products, the weather and costs that the company has to pay.

CONTRACT MILKERS AND SHAREMILKERS

The simplest way to think about it is that contract milkers get paid a contracted amount while sharemilkers get a share of the income.

Contract milkers are on contracts where they get a set amount of money per kilo of milk solids. They get paid to manage the farm but they're self-employed. From the money they're paid, they usually have to pay their employees' wages and cover the cost of other on-farm expenses like vehicles, power and contractors. What they're expected to pay for should all be covered in their contract with the farm owner.

Sharemilkers do much the same as contract milkers but instead of getting a price per kilo, they get a percentage of the total milk cheque. They can milk another farmer's cows, but they usually own their own cows and milk them on another farmer's land, then share the profit. If you're sharemilking, your income can vary depending on how much milk your cows produce and what price your dairy company is paying for milk solids.

ENTERING COMPETITIONS

There are heaps of competitions out there for young dairy farmers — the Dairy Industry Awards, the Ahuwhenua Awards and the Young Farmer of the Year are just a few of them. Enter them! They get you recognised, you get headhunted for positions and that propels you through the ranks really fast. They add value to your life and your business.

I went contract milking at 21 — 21! I was running a

business worth $7–8 million. My parents were wage-earners in Tauranga, and I had no family down here, so that shit only came from entering competitions. If I hadn't entered the Primary ITO Dairy Trainee of the Year or the Ahuwhenua Young Māori Farmer, I wouldn't be where I am right now. They've played a huge role in getting me to where I am.

So, if you see a competition — enter it! What's the worst thing that can happen? Nothing. That is the worst thing that can happen. If you don't win, you're still exactly where you were when you entered. You literally have nothing to lose. Only good things can happen. Pride stops people from attempting to be better because they're so scared of failing, but not winning doesn't make you a failure. That is success because you've had the balls to take a crack.

MOVING DAY

The dairy season goes from 1 June to 31 May, so 1 June is known as Moving Day (but some of you might know it as Gypsy Day). In terms of moving farms, it's really easy to wind up all the reels on the farm, pick up all your standards, move all your calf gear, move your trucks and tractors and get the stock sent over to your new place. What's actually overlooked is moving your staff, making sure that it's an easy shift for them and their families.

We need to make sure we communicate often so people understand their roles and responsibilities, especially when it comes to things that they might forget about — maybe cleaning the house once they've shifted their stuff out, getting insurance transferred over and making sure the power around the farm is sorted. The best way to do this is to get the whole team together for a meeting, printing out the moving day checklist from the DairyNZ website (and using it). Having a plan and these checklists will give you the confidence to make sure that everything that needs to be done has been.

ROAD RULES

In June, farmers aren't milking, so their cows will be grazing on feed paddocks, on roadsides — especially when it's been a bad growing year — and in places they wouldn't normally be, but behind one-wire electric fences.

Usually, cows will be grazing crops and they'll be behind a four- or five-wire fence as well as a few permanent fences protecting them from a main road.

When cows are out on the roadsides grazing, they don't have someone standing watching them to make sure they don't break out from behind the electric wire.

Our usual process is to put the cows down on the roadsides during the day, so we can keep an eye on them when it's light, then at the end of the day, those cows will be pulled back into a paddock.

With that all in mind, those cows are at risk of breaking the fence and getting on the road and being hit by a passing motorist. It's really important that road sites are checked regularly if you've got cows out there, and make sure your power is cranking through the wires protecting the cows. Check your road gates often to make sure they're closed and that they're in good condition.

If you live or travel through rural communities, keep in mind that there can be cows by and on the road during winter. Slow down, take time and take care to give those cows space when you see them out there. Even if you haven't seen them, but you know you're in a dairying area, it pays to take care as they can pop out from anywhere.

If you live in urban areas and you're travelling on rural roads, keep in mind that sheep, hawks, cows and horses can just fly out from anywhere.

FARM CONSULTANTS

Most dairy farm owners employ a farm consultant. Their role is to give the farmer, contract milker and/or sharemilkers independent advice on pretty much every aspect of farming. They're an important part of the team because it means you've always got someone to talk to if you need help with anything farm-related. That could be anything from herd planning to employment contracts to feed budgets to environmental issues. It's their job to keep up with what's happening in scientific developments, employment law, environmental regulations — you name it, they need to know about it so they can tell you what you need to do.

It's pretty important that you have a good working relationship with your consultant, and even more important that the farm owner does, as they're often the ones making the decision about whether to follow the consultant's advice or not. If you get on well with your consultant, it also means you're more likely to go to them when you need someone to talk to or get advice from.

FARM SOURCE

Fonterra is New Zealand's biggest dairy co-operative, which means the farmers who supply milk to the company are also its shareholders. The co-op also owns all the Farm Source rural supply stores around the country as well as running the Farm Source website, which aims to give dairy businesses support and information about the industry.

One of the best things about the website is that it has a really good jobs section. If you're thinking about getting into the industry or you're already in it but are looking for a new job, it's a really good place to start. As well as having links to DairyNZ's Go Dairy site, which has heaps of information about getting into the industry, the Farm Source Jobs page has a search engine where you can put in what sort of job you're looking for and the location where you want to work. When you hit 'search', a whole list of possible dairying jobs will come up giving you heaps of options all around the country.

When you go into a job listing, you'll find a job description, what accommodation is on offer, how many hours a week you'll be working and on what roster, and some of them will even tell you what pay they're offering. As well as that, there are details on the size of the farm, how many cows they're carrying and how many people you'll be working with.

Usually, the farmers will include a contact number or you have the option of applying for the job online. It's really easy to use and it's a good way of seeing what jobs are out there.

9.

WHERE THE GRASS IS GREENER

I went home that night, got on my computer and googled 'Southland'.

The first result was for a town called Winton. It had about 2000 people and three pubs — the top pub, the bottom pub and the middle pub. It also had a rugby team called Midlands. I thought, 'This could be the place for me. I'm not bad at rugby so I could come in here, be an enforcer and play some good rugby.'

Even though it was a really small little town, I thought it must have been the main town of Southland because it was the one in all the photos. I asked a few people about it, and one of them told me there were no shops and you had to buy your bread and milk from the pub. I didn't have any reason not to believe them!

Ian said that Craig's wife Rachel had a sister who lived down there. I went to see Rach and she told me about Southland and about her sister Glenda and her husband Graham Haynes.

The next step was to find a job down there. First of all, I went onto the Fonterra Farm Source Jobs page to see what kind of jobs were going. There were heaps of cool jobs — all in Southland! My goal was to get enough experience to become a farm manager then head back up home. I reckoned it would take about five years and would put me well on my way to eventually owning my own farm somewhere in the Bay.

I rang up a few places and they all said they wanted someone with rotary experience. I didn't even know what a rotary was. I'd always worked in herringbone sheds.

My uncle said to me, 'You should ring one of the advisors down there and ask if you can send your CV. They probably look after a few farms each.'

I found an email address for Ivan Lines, who was a farm consultant down in Southland. I sent him an email saying, 'Hi Ivan, my name's Tangaroa Walker. I'm an 18-year-old farmer up in Tauranga but I really want to move to Southland. I'm currently doing my level four with AgITO and I want to work for a family who's wanting to adopt a kid. I really want to learn about how to run a farm down there.' Then I attached my CV and pressed 'send'.

Ivan messaged me back that same afternoon. He said he had a couple called Wayne and Debbie Little, who were looking for a farmhand/2IC.

I thought, 'Woah! It's happening . . .' I couldn't believe it.

I phoned the Littles and Debbie answered. She told me all about the job they had going, and we had a bit of a yarn. Then she said they wanted me to start in two weeks' time. I said okay.

I went to milking the next morning and told Ian I'd got a job.

'That's awesome, when do you start?'

'They want me down there in a couple of weeks.'

'What the hell?'

Damn. The last thing I wanted to do was upset Ian. 'I don't know what to do. You told me to look and I looked and got one.'

The news soon sank in and Ian was really good about it. 'Nah, that's okay, we'll work something out.' Then he said something I'd heard him say many times over the years I'd worked for him: 'It's all right. No matter what happens, Tangaroa, I'll still be here milking my cows.' And I knew he meant it. No matter what happens with me, he'll always be there.

Soon, Ian got his head around the fact that I was leaving, and I talked to the Littles again, and they gave me another couple of weeks before I had to start work. Then it was just a matter of me getting everything ready to go down south.

I said to Dudley that I had no idea how I was going to get all my stuff down to the South Island and he goes, 'Well, you don't have enough stuff to bother taking a trailer with you. Why don't you make a box for the back of your truck?'

Dudley helped me make this big plywood box to go on the back of the Rodeo. It was about a metre and a half high and two metres long with a big door on the back of it. I stuck so much stuff in it that you couldn't have let a rat loose in it. I had my bed, which was in pieces, the mattress all rolled up, my washing machine, dryer and fridge, which were full of all my clothes, my dive gear, my stereo and speakers, you name it. I left the back of the car for the dog, Chase, and my two cats, Bessie and Blackie.

It was pretty hard leaving Mum. She was still going over to

Hamilton all the time for transfusions. She was getting really down and had no energy. Whenever I'd go over to see her, I'd take meat for her, and I'd give her money or go get 150 bucks' worth of groceries to fill her up for a while.

Whether or not I could leave my mum behind was my biggest decision. I knew that I was a motivation for my mum, and when I left she lost her support. I couldn't go and sit with her, I couldn't be there for her. When I left it felt almost like the death of my mum. I questioned myself a lot. Should I have done it? Yes, I should have. Could I have done anything more for her? Probably not. It was just the way it was. I didn't know how much longer she'd live, but I couldn't put my life on hold.

She told me that I needed to go, but I felt really selfish. I was trying to think about not only my career but the family I wanted to have one day and being able to put myself in a position where I could comfortably look after them.

I packed my dog and the cats up and went over to Te Puke and said goodbye to Mum and Gypsy. It was so frickin' sad. I knew it was probably the last time I'd get to spend a lot of time with her because I wouldn't be coming home for extended periods very often. My goal was to be in Southland for five years. I thought, 'Sweet, I'll be home by the time I'm 23.' That thought made it a little bit easier on me.

My last stop before I got on the road was at my auntie and uncle's place. I drove up there in the truck with all my stuff on it. I said goodbye to them then drove away. I knew they were both really proud of who I was and what I'd managed to do in my life.

As I went down the driveway, I was crying, and my uncle was bawling his eyes out. I used to do everything with my uncle, he'd be at all my trainings, all my games, my whole life. He was fully involved in my teams, my club and my life. His best mate was leaving.

As soon as I got on the road, my feelings changed. I suddenly felt like I was going on a big adventure. I was leaving to go somewhere I'd never been before. I owed nothing — I'd paid off my truck — and everything in front of me was all undiscovered. I had everything I owned with me. It was wicked.

I couldn't believe how long it was going to take me to get to Southland. If I drove non-stop, it was going to take me 32 hours. That was a bit longer than usual, but because I had the dog with me, I had to stop every few hours to let him out for a run. It was a good chance for the cats to get some fresh air as well. It was about seven hours to Wellington, then I had to wait to catch the ferry. From there, it was six hours from Picton to Christchurch, then about another eight hours from Christchurch to Winton.

I arrived in Wellington at about ten at night and the ferry wasn't until six in the morning. I decided to stop underneath the railway bridge by the ferry terminal and sleep in the truck. I didn't know how long I'd been asleep when a big ship's horn started honking really loud.

I was like, 'What the hell's that?' I jumped out of the truck and saw that the ferry was there. 'Shit! I've got to go.'

I jumped back into the truck. The loud noise had scared the animals. The dog had shat itself and the cats had pissed everywhere. The smell was so bad that I was dry-retching. I got the dog bed out and gave it a quick clean, then pulled the cat bed out and chucked it in a bin. I was sure we were going to miss the boat but there was no way I could drive with that stink going on.

Once I'd sorted that all out, I zoomed over to the bridge where you drive onto the ferry hoping I wasn't too late. I pulled up and gave the crew member my ferry ticket.

'You're not booked on this boat,' she said.

'Eh? What time is it?' I was confused.

'It's two-thirty. Your ferry's another three hours away.'

What! How the hell did I not check the time?

Then the lady from the ferry company said, 'It's okay, we can get you on this one if you want.'

I was so happy.

The ferry arrived in Picton at about 5.45 am. In less than 24 hours, I'd gone from being at Auntie and Uncle's in Whakamārama to being in Picton in the South Island — THE SOUTH ISLAND! I couldn't believe it.

It was October, so the sun was up by the time we got off the ferry. I went and filled my truck up and had a bit of a geez around there before heading down the coast to Kaikōura.

Eventually, I got to Canterbury and did a bit of a detour to Methven. My old mate Patrick was working down there while he was studying at Lincoln, so I went and stayed with him for the night. It was so good to see him again. He showed me around the farm he was working on. I'd never seen large-scale dairy farming like it before. I only knew the places where I'd worked up in the Bay and this was next level. These places had hundreds more cows, their sheds were huge, and the green of the plains seemed to go on forever.

Paddy was a big reason why I decided to move to the South Island. I figured if he could do it, so could I. Our other mate, Andrew, who used to work with us up in the Bay, was living just out of Methven, and some of the guys I'd been studying with at AgITO in Te Puke were down there, too. All the boys got together, and we had a barbecue, then went down to the Blue Pub.

The next morning, I put the dog and the cats back in the truck and we drove off south. Just after Palmerston on the way to Dunedin, I couldn't believe how steep the hills were that I was driving on. I felt like I'd been on a 30-degree gradient for about 20 minutes. It was unreal. I just wasn't used to driving on roads like that.

I dropped into Dunedin and caught up with some of the boys I'd gone to college with. They were at university there and it was frickin' cool to see them. They were telling me about all the rugby teams they were in and all the stuff they were getting up to. It was wicked.

They told me I was going to be living just a couple of hours' drive away. It was mean that they were so clsoe, but to me it was quite a long drive because it's the same as driving from Tauranga to Auckland — I always thought Auckland was a long way away! It wasn't until I moved to Southland that I realised a two-hour drive is really nothing.

From Dunedin, I drove through to Winton on the back road from Mataura, so I didn't go through Invercargill. The farm where I was going to be working was at Lochiel, which is about 10 minutes south of Winton, just off the main road to Invercargill. Wayne and Debbie Little were sharemilking 450 cows on a farm owned by Kathy and Lloyd McCallum.

I turned up at Wayne and Debbie's, and their three kids, George, Caitlyn and Daniel, all came out of the house to see me and they gave me big hugs. They were all really excited and I was happy to see them. I had dinner with the family and Debbie offered for me to stay the night with them. I didn't want to put them out so I said, 'Oh no, it's okay. I'm happy to just go over to the house.'

At the end of the night, they took me over to where I was going to be living. It was quite late and I'd had a long day, so I just pulled the mattress out of the back of the truck and chucked it on the ground. I slept really well that night.

The next morning, I was really curious to see what the farm was like. It was really cool to see all the calf sheds, and they had a big herringbone milking shed so that was good. It was a pretty cool experience for me. I thought, 'This place is mean!' Up north, I'd only ever milked in herringbone sheds. These are designed like a fish skeleton, so the cows all stand on angles in stalls on either side of a central wall. They can walk in themselves and then back out when milking is finished. The system was invented by a Waikato farmer in the 1950s, and it's now used on about 70 per cent of New Zealand farms.

Down in Southland, there are quite a lot of rotary sheds — and until I moved down, I'd never seen one. They were invented by a Taranaki farmer in the 1960s. In a rotary shed, the cows all walk onto a round platform in the middle that is turned by a motor. They are milked as the platform rotates then, once they're done, they reverse off once the cups are removed. Cows are always getting on and off the platform as it rotates making it a fast way of milking a lot of them in a shorter period of time.

It was October but it was really hot because they were having nor'westers.

I rang Uncle and told him all about the trip and where I was living, then I unpacked the truck. Once I'd got everything sorted, I decided to go for a drive to Invercargill to see what it looked like. I headed out the drive and down towards the main highway, then the frickin' box blew off my truck and broke into all these bits. I had to go and pick them up from all over the road.

After that, I decided not to go into town after all and I headed back to the farm to fix the box up. Later that night, I finally went on my mission into town. I couldn't believe how straight the road from Winton all the way to Invercargill was. There's maybe three slight bends in about 30 kilometres.

By the time I got into town it was dark. I got to the Lorneville roundabout on the outskirts of town and there were just all these lights in a massive long straight line next to huge powerlines that were going in the same direction. I drove from Lorneville into the centre of town and couldn't believe my eyes.

'Woah! This place is HUGE! They've got McDonald's and K-Fry! There's heaps of petrol stations and they've even got a Warehouse!'

I was absolutely blown away by how much stuff there was in Invercargill, because I'd thought that Winton was the biggest town in Southland. When your expectations are that low you can trip over them, it's so easy to get blown away — and I was.

I went and got some takeaways and some groceries, then went back to settle into life on the farm.

The following week, I signed up to start playing rugby for Midlands, the team in Winton. They were doing pre-season training, so that was a pretty cool way to ease into things. I literally didn't know anyone at all. I was so young and so happy to be away exploring the world.

At my first rugby training for Midlands, this guy rocked up in a Mustang. He jumped out of the car and he was wearing a Chiefs jersey. I thought, 'Jeez, this guy must be from the Waikato!' I was instantly drawn to him and thought he would be someone I could relate to, so I went over and shook his hand.

He said, 'How'z it gaan?' in a real Southland farmer drawl.

I said, 'Chee, bro, you from up Waikato way?'

'Nah, it's just my favourite team. You from up there, are you?'

'Yeah, I just moved down here.'

'Oh, yeah? What do you do for a job?' he asked in what I later found out was his typical cheeky way.

I told him I was working on a local dairy farm.

Blair was a hearty-as Southland farmer who just loved watching the Chiefs. He was a tall, lanky fella, but he was strong. He really reminded me of the top players I knew up in the Bay. He was a real hard rugby-playing farmer. Even though I didn't know him, I felt like I had a friend in Southland.

From then on, I really admired him as a person and looked up to him. He had this aura around him that really drew me in.

Then I met a guy called Razor — Ray Hemi. He said, 'Bro, do you play touch?'

'Yeah, bro, I'm the man at touch!' I told him.

A couple of days later, Razor came to the farm and picked me up then took me out to Drummond, which is about 20 minutes out of Winton on the way to Ōtautau, to play touch. That day, I met all these Māori boys. I think we just gravitated towards each other, because we all had the same values, we were all hori and none of us cared what we looked like and all the rest of it.

I sprained my ankle playing touch, so I limped off and went and sat down. One of the other guys, Shaun Harden, had also sprained his ankle. I said, 'Chur, brother, I'm Tangaroa.'

He was like, 'Chur, bro, what are you up to?'

'I've only just moved down here.'

'Oh, mean. Where are you from?'

I told him where I was from and he said, 'So what are you up to this weekend?'

'Sweet FA. What are we up to?' I said, being a bit cheeky.

He laughed and said, 'Is that us?'

'Too right, that's us!'

That Friday, Shaun messaged me: 'Come pick me up!'

I went and picked him up and that was how we became friends. We ended up going to this party down on Mavora Crescent in South Invercargill. I got really confused because there was Mavora Crescent, Mavora Place, Mavora Mews and Mavora Court all in this one weird little area. And I'd thought the pub names in Winton weren't very original!

At this party, I didn't know anyone except Shaun, but we got on anyway. I saw Sonny Rangitoheriri, who was playing for the Southland Stags. I was like, 'Chur, brother!' Then I saw Robbie Robinson — he was playing for the Highlanders. Unreal!

That night, I met pretty much all of the guys who are my really close friends in Southland now. It was unreal. It all came from Ray Hemi inviting me to play touch and from me asking Shaun what we were up to on the weekend. I had no worries in dropping the penny or sowing the seed of those friendships. People were inclusive and welcoming, which was so cool.

After that, I played touch in town and out at Drummond. I was playing for Midlands seniors as well. Then I heard there was a Southland Māori team about to be named for a tournament up north. I said I was keen for it and got selected. I also made it into the Southland sevens team. My boss was happy to give me time off to represent the province, which was awesome of him.

We went up to Timaru to play at a tournament up there. There was a whole busload of us went up to Timaru — the Southland Colts, the senior men's team and the Southland women's team. Some of the boys had driven up in their cars, but most of us were on the bus.

While I was there, I met up with heaps of the guys I'd played rugby with in the Bay, who were living in Canterbury and playing for their Māori team. I played pretty well in that tournament and got a couple of good tries. That gave me a bit of mana in the team and helped me to click with everyone a bit more.

We were meant to stay in Timaru to socialise with the other teams, but a couple of the boys said, 'Nah, man, let's go to O week in Dunedin.'

I had no idea what they were talking about, so they had to explain it was orientation week for all the students. My next question was, 'Is it any good?'

'Bro, it's mean as — let's go!'

It turned out that all the young Māori boys from Southland were going, and all of the older guys were staying in Timaru. I was 18, so it was a done deal.

When we got to Dunedin, we all met up at a friend's house to get ready. There, we met up with all the Bluff crew and a bunch of the other boys up from Invercargill.

My mates Monty and Tukes were there. They'd just been in trouble, so they weren't able to play at the tournament that weekend. They were gutted that they'd missed playing for

Southland Māori but decided to make up for it by having a big weekend in Dunedin.

Just seeing the potential walking around was pretty crazy. I thought, 'This place is mean!'

As well as being out with the boys, I caught up with heaps of my mates from school. We went to a couple of nightclubs and I was blown away by how good the party life was.

I ended up being wingman for one of the boys and we were walking up one of Dunedin's steep-as streets. Halfway up the hill, we saw some guys standing over another fella on the ground. We ran up to them and they took off.

The guy was just lying there on the ground not moving. I rolled him over and said, 'Hey, man, are you all right?'

'Yeeeaaaahhhh' he replied. He was wasted and he'd comaed out, and these guys had come along and stolen his wallet.

'Where are you from, bro?'

He groaned, 'I'm from Tauranga.'

'Oh, true, bro! Same. What's your name?' I said.

'My name's Smarty . . .'

'For real? Like David Smart's brother?' I couldn't believe it.

'Yeah, bro, do you know my brother?'

'Oath! He's one of the boys . . .'

I ended up helping him up and taking him back to his flat. It was a crack-up that he was there comaed out and I knew his brother. That's so Dunedin, though.

Once we dropped him off, we decided to head back down the street to catch the bus as it came through from Timaru. I'd never been to Dunners and the bro didn't really understand where the hell we were either. When we got down to the main street, we were meant to turn left but we turned right. We totally missed the bus, but it was worth it. It had been a mean-as trip. It was awesome to have played some pretty good rugby, met all the Southland boys and caught up with so many of my mates from back home.

Coming back to the farm after that trip, I really felt welcomed and like I'd settled a bit more. At the time, I was playing out at Midlands. It was a cool club, but while I was up in Timaru a couple of the guys were like, 'Bro, you should come and play for Marist!'

'Oh, why's that?'

'That's where we're all playing.'

And that was it. I switched from playing for Midlands to playing for Marist, so I could play with my mates. Their clubrooms were in the middle of Invercargill, so I used to drive the half-hour in from Lochiel a few times a week for practices and games. From then on, a big part of my life revolved around the rugby club.

10.

THE BEST MĀORI FARMER IN THE WORLD

After I'd been working for Wayne and Debbie for almost a year, Mum got really sick. She got a big infection inside because they'd put this needle in and pumped the blood back into her. Only the needle wasn't in the vein, so they pumped all of this blood into a cavity inside her body. She got really sick, so I flew up to see her. She started coming right, so I came back down to Lochiel. That was so hard as it felt like the last goodbye every time. I never knew if I'd see Mum again.

When I got back, I agreed to work for Wayne and Debbie for another year. They were amazing to work for. Wayne taught me all about efficiency. When I'd been working for Ian, he gave me all the time in the world to do my jobs, but he was running 290 head, whereas Wayne was running 450. The other difference was

Ian had three of us working for him, but in Lochiel there were only two labour units — me and Wayne. That meant that I had to find more efficient ways of doing things so that I could get all my jobs done in a day. It was a good lesson. He taught me about timeframes, and he explained why. 'We've got to get this done by this time because we've got to milk . . .' Things like that.

It was really good for me as it gave me a good structure to work around. I was on a 12 and two roster during calving, then 11 and three the rest of the time. That meant that I'd work for 11 or 12 days in a row, then have two or three days off. I'd work the morning on Saturdays when I was rostered on, so I could go and play rugby in the afternoons. One of the big differences I found in Southland was that here they get contractors in to do heaps of jobs around the farm, whereas up at Ian's we did everything — all the fencing, all the fertiliser, any digging we brought the Bobcat over and did it all. Here they'll just hire someone to come in and do it. The main reason is that a lot of the dairy farms down here are recent conversions and on a really big scale.

Once I'd got to know Wayne a bit better, I told him a bit about my story. I said, 'Have you seen *Once Were Warriors*?'

He said he had.

I told him, 'It was sort of like that. I saw heaps of alcohol abuse, drug abuse, verbal and physical abuse. Kids were getting hit all the time and it was Monday to Sunday drinking. There was heaps of poverty, kids not going to school because they had no shoes, that sort of thing.'

Wayne couldn't believe it. 'Nah, that's just a movie, T!'

'No, it's not. That's based on families up north,' I said. I was pretty surprised that he didn't think it was real.

'Naaah, it's all rubbish, that gang stuff, I reckon.'

He was quite a hard-out dude, and because he hadn't seen it for himself, he didn't believe me. That was until the Greazy Dogs came down here for the Burt Munro Challenge, which is a motorbike rally held every February out at Teretonga Park and Ōreti Beach. All my uncles and their mates came out to the farm to see me. About 40-odd Harleys rocked up to the farm, so they could see their nephew . . . hello! Who's lying now?

Wayne wasn't there at the time, but one of the farm owners was. He heard the growl of the bikes as they were coming up the driveway and made himself scarce pretty quickly. I forget sometimes that some people down here have had more sheltered lives than me!

As well as working and playing rugby, I was still studying and working for my apprenticeship. In 2010, AgITO opened up applications for a programme called 'Farming to Succeed'. The point of the programme was to identify future farming leaders and send them to do a bunch of workshops, visit farms and have discussion groups that focused on the business side of farming.

I decided to apply for it and was stoked when I was selected. They chose 12 young farmers from the South Island and 12 from the North Island to do a five-day programme run by Grant Taylor.

The course started on a Monday, so I decided to go pig hunting with Patrick and Andrew on the Saturday before. I left Invercargill in the early hours of Saturday morning and arrived in Christchurch a bit after seven. Even after driving all night, I was ready to go hunting, but all the boys were still asleep as they'd been on the piss the night before. I soon made them get their sorry arses out of bed!

We went over to Lees Valley to grab some hunting dogs from some of Patrick's family up there. Then we headed over to Mount Richardson near Oxford. At the start of the day, I looked way up to the top of this massive mountain and thought, 'Imagine if we walked all the way up there!' And we bloody well did. It was ridiculous. It turned out to have a public walking track right up it, and I thought to myself, 'What are we going to catch up here with paths with two hundred people on them? Nothing's going to be up here!'

We were walking up there with our pig dogs and there were people passing us, chatting away, every five minutes or so. There was no way we were going to get anything up there. Every now and then it would be quiet, and you'd see a bit of a sign that there were pigs around, then some townie out for a walk would pop out of the bush!

We walked so high up we could see the Lees Valley down the other side. Paddy pointed out where his family's farm was. We'd walked ages with no sign of anything but other people. Good one, boys.

By the time we got back down to the truck, we were so hungry. We hadn't eaten breakfast because I'd made the boys get ready real quick, then we'd gone to Lees Valley and walked up to the top of this bloody mountain — and we still had to take the dogs back.

It was about an hour and a half's drive on a windy road from Oxford over to Lees Valley. We stopped at a garage on the way. They had no hot pies left so we bought some frozen ones and we were that hungry we ate them. They were so nice!

It was about eight o'clock before we drove in to Rangi and Jack's place. I'd been asleep for a while when we got there. I woke up when Patrick let the dogs out. When he saw I was awake, Patrick said to me, 'Oh, you might want to stay in here. This guy doesn't really like Māoris.'

I was like, 'Oh, sweet, all good.' I stayed in the truck and went back to sleep.

I woke up a while later and the boys were still inside. They were taking ages. I decided to go in and see where they were. I walked into the house and that was when I met Jack and Rangi. They were the haatiest Māoris I'd ever even! I'm sure they were from Gizzy or Hawke's Bay or something.

Rangi came over to me straight away. She was just a real classic auntie. 'Kia ora, boy!' she said as she hugged and kissed me.

Then Jack came over. 'Kia ora, bro,' he said in his big, deep voice. Then he held out one of his massive hands for me to shake.

I just burst out laughing. 'Chur, brother! These pricks here told me that you fellas didn't like Māoris!'

In proper auntie mode, Rangi said, 'Are you guys hungry?'

'Oh, a little bit . . .' we lied.

Rangi went off into the kitchen while we had a couple of beers with Jack. Next thing, she came out with this big roast meal with mutton and pork chops, rīwai and carrots. She poured this cream of chicken gravy over the potatoes. Oh my god! After being wet, cold and hungry and then getting served that feed, I felt like I was in heaven.

We smashed the feed because we knew it was going to take us another hour and a half to get back out to Oxford and we were going to the races the next day.

While we were talking, Jack suddenly said, 'Did you guys get something?'

We were a bit embarrassed. 'Nah . . .'

He looked at Patrick. 'So your mate drove all the way up here to catch nothing?'

I said, 'Yeah, these pricks have been talking this place up. They reckon there's heaps of pigs up here, and I get up and hello, nothing!'

Jack said, 'Grab the dogs then. Chuck them on the truck, you fellas. Go get something.'

It was about 10.30 at night and I was not that keen. Because he was haati as, I said, 'Yeah, bro, yeah, sweet.'

We took the dogs out into the bush around Lees Valley and they disappeared. We couldn't just leave them there, so we kept walking trying to find them. Every now and then, we could hear them, but we couldn't work out where they were. After about four hours, we found the gully they were up, so we parked at the bottom. We hadn't planned on doing a night hunt, so all we had with us were our cellphone torches.

We were walking up this creek, through gorse, with just our cellphone lights to see by. We ended up finding the dogs and they'd rounded up this big-arse black boar. By the time we got there, they'd had it cornered for a while so it was pretty knackered, which was good because we were, too. We got it, but then we had to carry it back and it was huge.

By the time we got back, it was about five in the morning and we still had to drive out. By that time, I'd been up for about 28 hours solid.

We drove out and the next day was the start of this 'Farming to Succeed' course. I rocked up there absolutely knackered. It was a five-day course and I thought it wouldn't be that bad. Our days were unreal, but it was really full on.

Every day, we'd start at 7.30 in the morning and we'd finish at about midnight. Everyone was just so engaged and so keen to listen and learn about how to hustle business. It was really cool being with other young farmers from all over the country who were working in all parts of the industry.

As part of the programme we visited all these farms, starting with Kathryn and Leo van den Beuken's dairy farm about half an hour out of Ashburton. We also went to a feed lot and some sheep farms. In classes, we learned a lot about progressing in the industry.

It was such an inspirational week that I came back with all these ideas about what I wanted to do. It turned out to be the second-best event I've been to in my life in terms of goal-setting and action.

At the end of that season, I decided to stay on with Wayne and Debbie for another year. That was when I entered the competition to be named Primary ITO Dairy Trainee of the Year.

In 2012, the judging for the Primary ITO Dairy Trainee of the Year was held and I won it. That gave me my first experience of talking on the radio and getting interviewed for newspapers. It also gave me the opportunity to do a lot of networking. That resulted in me getting my first invitations to be the keynote speaker at events. Having been the captain of my rugby teams, I was used to public speaking, so luckily it was something that was quite natural to me.

Around that same time, a couple of people said I should enter the BNZ Māori Excellence in Farming Awards where they were awarding the Ahuwhenua Young Māori Farmer of the Year for the first time. As it was the inaugural year for the award, I didn't really know what to expect. The goal of the award was to encourage young Māori into leadership roles and to support them in their careers.

I filled out the application form and submitted my board of goals, which was designed like a Monopoly board. On every square I had a goal I wanted to achieve, and each side of the board was a year, so it laid out my four-year plan. Every time it showed going through a bit of risk, I put an arrow pointing towards the square that had 'Jail' on it.

I found out I'd made the top 10 for the Ahuwhenua Young Māori Farmer of the Year. From there, I had to do some online interviews. During this whole process of applying for the Ahuwhenua Young Māori Farmer Award then getting into the top 10, I found out that Uncle Kelly's dad had tried out for the Ahuwhenua Trophy for dairy farming. That gave me so much more motivation to have a really good crack at it.

In the meantime, I was coming up to the end of the season at Wayne and Debbie's. They knew that I'd outgrown their business, so they encouraged me to find a more advanced role. I ended up applying for a job managing a farm for Rachel Jeffrey's sister, Glenda, and her husband Graham, who were contract milking on quite a few properties.

They were looking for someone to work on a farm they were contract milking for an equity partnership, Toa Farms. The company had quite a few investors, including Murray Hewitson and Wayne Clarke. When I went for the interview, I told Graham and Glenda I wanted to be contract milking in a year's time. I got the job.

The cool thing about them was that I knew I could relate to them. Graham was a lot like Uncle Kelly. He was into rugby, he was a Tauranga sportsman himself, and his son had played for Te Puna. I couldn't have had a better connection.

I started working for them at a farm out at Rimu, which is just off State Highway 1 east of Invercargill. I went from living in Lochiel to living in Kennington, which was cool as it was a lot closer to town for rugby. Instead of taking me half an hour to get into town, it took about 10 minutes to get from home to Marist. Around that time, I was still playing for Marist and the Southland Māori sevens team. I made the South Island Māori team, but I never played for the Southland sevens team because I couldn't make it to trainings.

I was 20 years old and managing a 570-cow farm. I also had a guy called Gordy working under me as a 2IC. Gordy was in his late forties, so the dynamic of having someone working for me who was twice my age was really interesting. I didn't like giving him all the hard jobs, so we shared them out. He was a really hard worker, and we got on pretty well.

From Ian, I'd learned how to do jobs around the farm, and from Wayne, I'd learned about efficiency, but I still didn't know much about numbers and feeding cows properly. Those were things I learned about, working for Graham and Glenda.

I had only been working for them for a few days when I found out I'd made the top three of the Young Māori Farmer of the Year. The three finalists were all interviewed on their farms, so the judges came down to Rimu. The judges that year were Fred Hardy, Abe Seymour and John Troutbeck.

Abe asked me, 'Why should we pick you?'

Straight away, I said, 'Because I'm the best Māori farmer in the world!'

He said, 'Well, you're actually the second best because I'm the best . . .' then cracked up.

The other two finalists were Tyson Kelly, from Te Awamutu, and Mark Coughlan from Mangakino. We spent three days together in Hamilton before the big awards night, where we visited a couple of farms, spent time with industry leaders and talked about our Māoritanga and our values.

After that, we went up to Auckland for the awards ceremony, which was held at SkyCity Convention Centre. Uncle Kelly and Auntie Peata were there, but Mum couldn't make it because she was too sick. My real mum and dad — Nancy and Hanu — my uncles and some of my cousins from my Ngāti Pūkenga whānau from Welcome Bay were there, too, because some of them were shareholders in a farm that was entered in the main award.

The whole of the convention centre was all done out in red and it looked mean. They had a kapa haka group do a pōwhiri to welcome everyone in. There must have been about 500 people there all dressed up in tuxedos, suits and nice dresses.

It was pretty unreal when they announced my name as the winner of the award. Uncle was in tears, he was so proud of me. He got up and did the haka to me with the rest of the family. That was something I never thought I'd see him do. It was just so moving.

I was presented with a trophy, a certificate awarding me a training scholarship and $3000 cash. Then I had to do a speech and I thanked everyone for the nomination and the chance to take Māori farming to the world. I said, 'The industry is screaming out for a young Māori farmer to teach people about why we farm and how we farm. I don't think I'm the right person for the job, but somebody needs to do it.'

That's something that's stuck with me right until now. Even though I didn't think so at the time, I actually was the right person for the job, it just took me a while to realise it.

When they asked the judges why I won, Abe grabbed the mike and said, 'Well, I asked this boy one question. I said to this boy, "Why do you deserve to win this competition . . .?"' Everyone was sitting there cracking up. He continued, 'And what did you say, boy?'

I was standing there going, 'I'm not going to say it!'

So Abe continued, 'This tall, dark, skinny Māori boy goes, "It's because I'm the best Māori farmer in the world!" And that's why you were the winner.'

Everyone was cracking up and I was cracking up on stage. Abe laughed, then said, 'There's one problem with us Māoris is we're too bloody humble.'

I got so much shit for that at rugby training the following Tuesday after they reported it in the *Southland Times*. 'Here he is! The best Māori farmer in the world!'

A bit after the awards, Uncle told me that a bunch of them had got together to practise the haka in case I won.

While they were there, Hanu said, 'Ah, what are we doing this haka for anyway?'

Uncle Kelly said, 'What do you mean why are we doing it?'

Hanu replied, 'Ah, he's not going to bloody win it anyway . . .'

Uncle reckoned he nearly smacked him for that.

What a weird thing to say. I was so dark. I couldn't understand how he could be like that.

Mum and Gypsy were gutted they couldn't be there, but they were really proud of me. Mum was starting to get really sick now, with complications from the line that went into her heart. She had to have operations, but they wouldn't heal, and she'd get infections.

I started going up to Te Puke every couple of months to spend time with her because she was so bad. Even with that, I still spent a lot of time feeling like it was touch and go whether I'd need to go back up home to spend time with her.

Graham and Glenda were really understanding of my situation and were very supportive in letting me have time off to go see Mum. We used to call Glenda 'Mother Nature' because you don't mess with Mother Nature! She was hard-out. Graham was a real numbers man. He could look at a paddock and go, 'There's X amount of grass in here. How many cows are going in here?' then he'd be able to work out the feed rate and tell me what to add into the paddock.

He started giving me a bunch of equations that I had to do every day. Since then, I've gotten way better at numbers. Now I go to a paddock and figure out how much grass the cows in it have been eating in the last 24 hours and do a rough summary of how much in the way of milk solids the cows will produce from it.

Graham also taught me all about systems and gave me a lot of rope to work out grazing plans and fertiliser rates. He gave me a protocol list of what to do if anything broke down and I would follow that to get stuff fixed. I had a lot of independence and responsibility, but I was answerable to him.

When we had our meetings, I'd turn up with all the info and he'd go through it all with me along with the shareholders and the farm consultant. Engaging me in those meetings really helped me get a clear view of the bigger picture rather than just the targets and where the cows needed to be.

As well as working hard for Graham, I was getting asked to do a bit of guest speaking. One of the keynote speeches I gave was at the Federation of Māori Authorities hui at Miraka, which is a Māori-owned dairy company, in Mōkai, just out of Taupō.

In my speech there, I expanded on what I'd said at the Ahuwhenua awards. I said that somebody needed to sexy farming up. 'Whoever that person is needs to still be farming, but they also need to have a vested interest in portraying farming for everything that it is, and not for everything that it isn't. I want to be that person.'

At the time, there had been a bit of negative press about dairy farming and I reckoned that it needed to be a farmer telling those stories directly rather than through the media. I told the people at that meeting that by telling the stories, we'd get more young Kiwis into the business and we'd get more young Māori wanting to farm their whenua.

After my talk, a lady called Ave came up to me and she was crying. She said, 'We own a farm at Matakana Island, and my son used to work with us there. He's now an engineer but I really want him to come back farming. Would you consider having a talk to him?'

I said, 'Oh, yeah, what's his name?'

'His name's Ronnie, Ronnie Gardiner.'

I ended up giving Ronnie a phone call and having a talk to him. I told him about my situation and asked him what he was up to. He sounded like a pretty onto-it fella, so I caught up with him the next time I was up in Tauranga. We clicked immediately.

Ronnie was into everything. He was an engineer, he used to look after the stock on the farm for his old man; he was a contractor; he used to sell watermelon on the side of the road, he even used to sell sweetcorn from Matakana Island up to the markets in Auckland. He was a bit of a hustler and a real out-of-the-box kind of a guy. He sounded like me!

I said to him, 'I might have a 2IC job coming up. It's not very good money but I'd be able to give you as much knowledge as you can take, and I'll answer all your questions.'

He was really keen to get out of Tauranga, so Ronnie and Makuini sold their house to come down to Southland to work for me. I felt privileged to be able to employ someone of that calibre based on the amount of money I could afford to pay him.

I also started work on my diploma in agribusiness using the scholarship that I'd won, as well as doing a bit of work in a business I'd started with my sister the previous year. She was married to a guy from Wyndham, and they'd come back from Aussie where she'd been working as a chef. She was unemployed, so we started a small business called 'Mr T for Tea', which delivered home-cooked meals. We had about 50 farmers and farm workers on our books, who we'd drop meals off to because they didn't have time to cook. My sister did a lot of the work for it, and I helped out when I could.

After that year with Graham and Glenda, they wanted to stop contract milking on the farm at Rimu. While that meant that I would be unemployed, it also meant there was an opportunity for me to take over the contract milking job on that farm. It couldn't have worked out better for me really.

I got the job, and the transition from Graham and Glenda to

me was quite smooth. I got a loan from ANZ Bank, who were offering finance for people who were starting up as contract milkers. I think I got a $30,000 loan and a $30,000 overdraft for the season. I was 22.

11.

DO THE MAHI, GET THE CREAM

Even though I was young, I really wanted to step up to contract milking because I wanted the opportunity to be the boss that I wanted to have, and I wanted to be the boss I thought I could be to new people coming into the industry. I also wanted to put a multicultural spin on business.

So when I was given the opportunity to take over the contract milking position on the farm I'd been managing, the farm owners had seen me manage the farm for a year and they knew Graham and Glenda were going to hold my hand through my first year of contract milking. It was an awesome year and I learned a lot.

I hadn't grown up in a family that knew about being self-employed, so going into that level of debt and employing other people was a big learning curve for me. Taking that financial jump was massive. It was quite a wet farm, but it was beautiful, clean and easy to run, and with good infrastructure. That made my life a lot easier.

Jack Ballam put together a contract, which really worked for me in my first year of contract milking. It set me up in a lower-order contract milking position, so I got a small percentage of the milk cheque and I also got a set contract amount every month.

Through the winter months in Southland, you're pretty much not getting any money as there's no milk going out because the cows dry off before calving. Even though you're not getting any money, you've still got to pay your staff, run your vehicles and feed yourself. The way Jack worked my contract, it included an estimate of the milk we'd produce for a year, then split that into payments over 12 months. That made it easier for me to manage my finances and not get further into debt.

I absolutely loved being a contract milker. I was just chipping through life really. I was doing some courses and still playing rugby for the Marist premiership team as well as the Marist touch team. It was about this time that I met my good mate Dan. His mum had been our calf rearer when I was working at Lochiel. She kept talking about her boy, who she reckoned I'd get on really well with. I'd never met him, but she kept telling me about him. I didn't really take much notice of it.

When I started playing rugby for Marist, this fella Daniel came along. We started playing rugby together and became quite good mates. It was only then I found out who his mum was and that he was the Dan I'd been hearing about. We found out that we both liked pig hunting, fishing and diving, and we're both dairy farmers. His mum was right!

I realised he was a lot like me — we're both hard-out kind of characters with plenty of energy and neither of us like to back down.

That led us to another, less pleasant bonding experience when five guys ganged up on us in the toilets of a bar in Invercargill. Neither of us are fighters, but that night we didn't have any choice. Me and Dan both managed to get out in one piece but the same can't be said for the bathroom cubicles. Luckily for us, the bouncer opened the door and said, 'You guys might want to shoot off . . .' to me and Dan. We walked out and he told us that the guys had been causing trouble all night. After that night, we knew we had each other's backs.

Rugby aside, I was really focused on the business. The only thing that was a bit sad was that, as a beginning contract milker, the bosses weren't keen on me running other businesses, so I had to pull out of doing the meal deliveries with my sister.

I wanted to rear some beef calves for a bit of extra cash, and Rex Crosland up the road in Rimu let me use his shed. I kept all my bulls in there. After weaning, I leased a small block in town where I took the calves to graze until they were up to 100 kilos. Once they made weight, I sold them. That meant there was a good bit of money coming in to pay off the overdraft.

After a year of contract milking, I was thinking about my next season on the farm. I had things on track, and I thought I was the mantis. I really thought I could take on anything in front of me. On the work front, I felt like I'd learned everything I could from it, like I'd got too good, so I went out and looked at new jobs.

Instead of staying in my comfy contract milking role and leveraging it to create wealth somewhere else, I said goodbye to this awesome farm with its really supportive owners and struck out for new pastures.

I got offered a huge job on a pretty run-down farm, and I really thought I could make a big difference by improving the place. The farm consultant told me that the new farm's production wasn't great, so I'd be able to really make changes out there while making a lot of money.

The farm owner came to the farm at Rimu to meet me. He knew that I'd won the Ahuwhenua and he said the guy he had working for him at the time wasn't doing the job properly. The owner sounded like a real hard-out guy. He talked really fast and made heaps of promises. He seemed keen for me to go and work for him.

I took the job because there was the potential to go sharemilking and the option to rear 70 heifer replacements. I knew if I could do that, I could sell them or take them up north and go sharemilking in Waikato. Another reason I took it was because the farm consultant, who I liked and trusted, was going to go with me. Together, I felt like we could really lift the place and make it crank. I was pretty excited.

I wasn't sure I had enough experience for the job, but having Ronnie working with me helped give me the confidence to make the move to this bigger farm. I also brought one of my nephews, Connor, down from Te Puna to work for me. Connor was a short, solid, strong little Māori boy. He was about 16 when he arrived.

Together, we were all working on this big property with 980 cows and a 50-bail rotary shed, as well as a 300-hectare run-off block out by the beach. I also employed a guy, Patrick, to work at the run-off, and Blair worked as cowshed manager on the main farm. It would have been a $30 million-plus operation I was running at 23. It was all a huge shock to the system. I ended up having to employ six staff, most of whom were older than me. If I'm honest, I was a bit out of my depth.

As well as going from 540 cows to 980, I went from running 210 hectares to running more than 500 hectares. Managing the new farm also involved wintering all of the stock on the farm, cutting and carrying all of our feed from a run-off property, drying off stock and rearing all the ones and twos.

At the second farm consultant meeting, the consultant had said to the farm owner that we needed to change a lot of stuff. The farm owner wasn't keen to take his advice, and the consultant basically said that if he didn't listen, he'd leave and take me with him. This was three or four weeks out from calving and I'd already signed my contract and employed all these people, so I had no choice but to get on with it. That meeting set the tone for the next 22 months for me.

The farm owner's main problem with what the consultant had suggested was that he didn't want to drop his cow numbers down from 980 to about 800. It was a good idea as the farm's per-cow production at that point was really low.

The other recommendation was that all the calving got done on the dairy farm instead of down at the run-off. At that point, the game plan was to calve all 980 cows at our run-off, which was 20 minutes down the road. Again, the owner said no. We had to go ahead with calving down at the run-off and we then had to pick all the calves up and cart them back to the dairy farm on the trailer.

Meanwhile, all the cows that had just calved were meant to be put on a truck and driven back to the dairy farm for milking.

We were getting about 50 or 60 cows calve a day, so I had to move them and their calves back every single day. I built this big trailer and we had to take it 20 minutes there, and 20 minutes back twice a day.

The truck never seemed to turn up when we needed it, and there'd be cows that had just given birth being loaded up onto the top deck of a truck. It was really rough on them, especially if they'd had hard calvings. To make it even worse, some of them would pick up mud and shit off the truck and get mastitis. We'd then have to milk them that night.

Because we only had a 50-stand cowshed and 980 cows, we ended up having to do 20 rounds of milking at the peak of milking. We were getting cups on at 3.30 in the morning and we wouldn't finish milking until about 8.30.

After that, we'd bring the cows off the truck and because they were getting dropped off at the shed, we'd milk them straight away. Usually, we'd milk the ones that had just had calves last, but we couldn't because of the time of day. That meant we had to put a wash through the shed before the next lot of cows came through. Then we'd pretty much start up milking again at 1 pm so that the boys could be home by 6.30–7 pm. It was a really backwards system and I could see why Jack had been so against it.

It was really frustrating as I'd been brought in to increase production, increase profitability and reduce the amount of expenses on the farm, but I couldn't do any of it. The farm owner just didn't want to change anything he'd been doing. I think he thought that the more cows you had, the more money you'd make, and it just doesn't work like that.

The unfortunate thing was that I wasn't getting paid that much to do the job, and I hadn't expected all of this to be going on. Instead of letting us get on with what we were there to do, he used to come and tell us to clean the shed or do chores. I stood up to him and he didn't like it. It was hard to believe that he'd employed my company to run the place, but when I got there he wouldn't let me do the job. I'd been told that I'd be lifting production while reducing the workload, and I could easily have done that if I'd just been left to do what I'd been hired to do.

The farm owner did all the cropping and the silage down at the run-off, so if he needed something like a tractor down there, he'd just take it, no matter if we needed it to put fertiliser on the dairy farm that was making all the money. The tail ends up wagging the dog in situations like that.

As well as making me lose confidence, it really affected my business as I had to pay my staff way more than we'd budgeted for because of the huge amount of work we had on, which we hadn't been told about. That was made worse by the fact I was working for a farmer who wasn't supportive of me or the way the consultant told him to run the farm.

The houses on the place were terrible, too. The farmhouse was this big, old homestead and it was the scariest house I'd ever lived in. It had high ceilings and huge rooms. There was a fireplace in the main room and a heat pump, but it was still absolutely freezing. I reckon it was definitely haunted, it just had a really weird, creepy feel to it.

Blair and Ronnie's houses were old, freezing cold and mouldy. The ancient wallpaper was peeling off, the carpets were all threadbare and there were cupboard doors missing in the kitchens.

They were pretty much derelict, and they expected people to live in there. It was no wonder they'd had trouble attracting staff.

The pressures on me were huge. I had people working for me, but I couldn't pay them what they were worth because the hours they were doing were way worse than I'd planned. I'm all about efficiency, people getting home to their families and people getting paid what they deserve, so the whole thing went against everything that I believed in. I hated that I couldn't look after my mates. It was the hardest thing that I've ever had to do in business.

I'd signed a two-year contract on the place and by the end of the first season, I'd had a falling-out with the farm owner because he just wouldn't listen to anything I said. He wouldn't listen to the farm consultants either, so while I was there we went through three farm consultants for that very reason.

By the end of my time there, I probably was doing a bad job because I wasn't allowed to make any changes or do anything to improve the place. I was doing the bare minimum because there was no motivation to try to do a good job. It was just such a struggle.

DO YOUR HOMEWORK

Don't get too cocky.

If you're on a good farm, stay there. If you have good owners who you trust, stick with them. Trust is the biggest thing. If you trust the people you work for and the people you work with, then that's the most important thing.

If you meet someone and you get a bit of a weird feeling about them, trust that feeling. If a job seems a bit too good to be true, then it probably is. Trust your gut and don't ignore any red flags you might notice.

Employees have to supply references, but I reckon employers should provide them, too. One way to get around that is contacting the people who were in the job before you. If you're looking at taking on a new role, whether it's a step up into management or just a better-paying job, make sure you find out as much as you can before you agree to a new contract.

When I was going into interviews with employers, I wish I'd rung up and asked the ex-employees about their situations, whether the owners of the farm were good, whether the manager was good to work for, why they left, were there any issues I should be aware of, that sort of thing.

The past employees have nothing to lose and have already moved on, so they shouldn't have any reason to lie to you. Ask them about the bosses, ask them about infrastructure, ask them about animal health issues.

Check the houses that you and your staff (if you have staff) are going to be living in. Make sure they're up to standard and liveable. Make sure they're insulated. Make sure there are no rats. Make sure the carpet is intact.

If you can, take your mother or your mother-in-law with you to have a look and ask her if she would live in those houses, or if she'd let her children live in those houses. She'll bloody tell you — and she'll tell the farmer if she gets a chance! If she says no, then either don't take the job, or tell the farmer that you're not going to move into the house until the standard of it has been improved.

It doesn't matter how good a boss you are, how awesome the job is, or how excellent your work team is, if your staff are going to nasty cold, damp houses they're going to be miserable people to work with. I learned that one the hard way.

I was having a real shit of a year, but one good thing did happen. One weird, random night, I said to Ronnie, 'Shall we go out?'

He said, 'Yeah, go on then.'

I picked him up and we ended up going into town really early, like 9 pm. In Southland, no one goes out night-clubbing until about midnight, which is ridiculous because you've only got three hours to dance and have fun. Up in Tauranga, we all would go out at about 8.30 pm so we had the whole night to drink and get up to mischief.

Anyway, this night, me and Ronnie went to the Lone Star ('where dreams are made!') and got on the piss. We were buying heaps of drinks because we go large when we go out. Then Courtney and her mate Jessie rocked in.

Courtney Heke-McColgan played on the same Marist touch team as me on Wednesday nights. The first time I saw her I thought she was pretty hot, but I didn't tell her. I would try to chat her up during games and she wasn't having a bar of it as she was so focused on playing the game.

One day, I said, 'How was your day?'

She actually replied. 'Yeah, good, how was yours?'

Yes! 'Oh, cool, I went out diving.'

'True? Did you get anything'

This was going well. 'We got some pāua. Can you cook pāua?'

She looked a bit shocked. 'Yeah . . .'

'You'll have to cook me some sometime.'

Yeah, that didn't work at all!

By the time me and Ronnie ran into Courtney and Jessie in town, we'd been on the same touch team for about six weeks. They'd been on a bus trip out to Ōtautau for Jess's work Christmas party, so they were absolutely wasted. I was quietly really stoked that Courtney was there.

We had a few drinks together, and while we were talking I asked Court where she worked. She told me she was at Craigs Investment Partners.

Later on that night, we were all dancing and I saw that Courtney was getting chatted up by this other guy. She said he was just a friend, but he was definitely chatting her up!

While she was talking to him, Justin Timberlake came on and the song was 'Señorita'. I thought, 'Here's my chance' and I went over and cut this guy's lunch.

'Chur, bro,' I said with a raise of my eyebrows.

'Hey, mate,' he said in a way that meant I definitely wasn't his mate.

'I'm sorry,' I slurred. 'I've just got to ask this girl if I can have a dance . . .' Then I started singing her 'Señorita' while I pulled out all my best moves. I was trying to be a salesman — 'You need this product! You want this product!'

She obviously decided to buy what I was selling because we had a quiet passionate kiss in the middle of the Lone Star dance floor. Then she looked at me and said, 'Don't you have a girlfriend?'

Shit. 'I'll get back to you on that one.'

Not long after that, I realised the time and we had to leave because we were milking at 3.30. Me and Ronnie got a taxi back to the farm, which cost us about $120, then went straight down and got the cows in.

Tuesday came rolling around and I decided to ring Courtney at work. It would have been Monday, but I needed the time to build up the confidence to make the call — and to google what Craigs Investment Partners actually did.

I rang the number and she answered. 'Hello, Courtney speaking . . .'

'Oh, hey, I'm just ringing up to get a little bit of advice about an investment that I bumped into on the weekend.'

She was like, 'Ummm, okay, what's the investment?'

I said, 'Oh, just this investment that I heard about over the weekend.'

'Okay, I'll just put you onto one of our advisors.'

Next minute, I heard 'Beeeep . . .' and I hung up straight away.

I found out later she hadn't recognised my voice so she had no idea it was me ringing up to be cheeky!

I rang her back pretty much straight away and said, 'Oi, it's me, you egg! What are you up to?'

She whispered, 'Oh, hi, I'm actually at work at the moment.'

Being a farmer, I didn't really understand. 'Can you not speak?'

Her voice got even quieter, 'I'm sitting next to other people. I'll call you back.'

After that I expected the worst and I was sure she wasn't going to call back. But she did.

I asked her if she wanted to go and catch up out at Ōreti Beach after work one night.

I didn't want to take her out for dinner in case anyone saw us. Small towns! You can't do anything down here without anyone seeing you and telling everyone. The beach seemed like a good idea because it's a huge stretch of sand and I knew I'd be able to find a place where there were no cars.

That day, I rang her and asked her if she wanted anything to eat as I was going to get some Subway. She said okay.

I was blown away that she said yes. 'Okay, what do you want?' and she reeled off her order.

I couldn't believe it. Most women would be like, 'Oh, no, it's okay, I'll just have a salad.' Not Court, she ordered extra cheese and extra sauce. What a good wahine!

I rocked up to meet her and she'd just got back from the gym. She kept rubbing her biceps, which I thought was a bit weird, so I asked her about it.

'We just did pull-ups tonight . . .'

Faaar, this chick is hard-out.

We talked a bit about our work and stuff, and then she told me about how she was into CrossFit and paleo diets and all that.

I was only half listening, and as soon as I got a chance I said, 'I really like you, eh?'

She said, 'Thanks.'

I thought 'Sweet' and leaned in for a kiss.

'Don't you have a girlfriend?'

I stalled for time. 'Ummm . . .'

Court wasn't having a bar of it. 'Nah!'

That was that. I drove back into town and dropped her home.

As soon as I got the chance, I was on the phone to Uncle to tell him I'd met this beautiful young lady from down here.

'But, Uncle, not only is she beautiful and intelligent, the killer for me is that she's a real boss as well. She can dance and she's a good singer AND she goes down to the islands and kills muttonbirds.'

I couldn't believe she liked tītī let alone killed them. She'd told me all about going muttonbirding, and when I asked her how many she did, she was like 'Hundreds!'

Up until then I'd thought she was a pretty little thing who knew what she wanted, but when I found out she was quite happy to get her hands dirty, I realised she was the perfect package. Then there was the added bonus of getting tītī!

Uncle laughed and said, 'Oh, well, if that's where your heart takes you then that's where it takes you.'

I rang Courtney the next day and said, 'I'm single now.'

It was pretty ruthless, to be honest, but I'd just fully fallen in love with her. It was weird because I barely even knew her, but I knew I was in love.

Court had grown up in Invercargill, then gone to uni in Dunedin to study finance and economics, then came back to do a couple of years' work experience before heading overseas. That all changed.

It's so weird when I think about it because she wasn't supposed to be out that night and it was pretty random that Ronnie and me had gone into town that night, too. Even weirder was that Jessie's work do was out at the pub in Ōtautau, which was the pub I usually drank at. The stars had just aligned.

It turned out our paths had crossed before, though. A while

later, I was looking at the booklet from when Court graduated at Otago. I'd done a keynote speech for one of the graduation ceremonies. Her name was on one page of the book and mine was right there on the page opposite it.

Courtney came into my life in December 2013. From the moment we got together, she used to come out to the farm all the time. She hadn't really been on a dairy farm before, and I was really impressed that she was happy to come out and just do stuff with us.

She was a big light in my life at that time when I was going through so much on the farm. It wasn't long before she got me thinking about things outside the farm. I started going to a CrossFit gym in town and working out with her. She was always just so positive it really helped me to see things differently.

On 16 January 2014, a couple of my staff were away on their Christmas holidays, so I was milking when I got a phone call at about 5 am. It was about Mum. My auntie rang to tell me that she'd died.

Obviously, I was gutted, but at the same time I was glad she didn't have to put up with the suffering anymore. She'd had a horrible quality of life for quite a while, and the last time I'd talked to her she'd told me that she was just over it. I knew that she'd gone to a better place and she was away from all that pain.

I had to get the boys in to look after the farm, so I could head up north to sort out the funeral. It was the first time I'd ever had to organise a tangi then go through the whole process of picking coffins and headstones and having to pay for everything as well. Because I had to organise the business and get flights, I didn't get up home until the day after she died.

When I got there, all my aunties had already taken everything out of Mum's house. They'd sorted all her stuff out and got rid of a lot of it. I was so dark about how that all went down. For me, that was part of the grieving process that I never got to experience. I understand that she was their sister, but she was my mum. It hurt that they didn't think about it enough to realise that I'd want to be there.

Gypsy was gutted. He was the one who had been with Mum all that time. My aunties didn't really like him, and I think they blamed a lot of her poor health and her death on him, but they didn't know what Mum was like. It seemed like they thought Gypsy was bad and that Mum was good, but Mum used to make bad decisions around food and alcohol herself.

I really felt for Gypsy because my aunties went through the house, which was his house as well, and cleaned everything out. All her clothing, all the photos, everything was gone. Now he lives in a shed at the back of a kiwifruit orchard. It's really sad. I've been trying to get him down here for years, but he seems happy where he is.

Missing out on that process meant I just didn't get the grieving that I needed. I still haven't cried about it, because she was there then everything was gone apart from my memories. I can't even go and hold something she wore or go to the house she lived in. It was a terrible part of that process for me.

The one thing that really triggers that grief in me is when I have mince and mash because that's something Mum used to cook for me all the time. I loved it. When she would ask me what I wanted to eat on my birthday, I'd always ask for mince, mashed spuds and apple pie. That's still my favourite meal and it really takes me back.

After Mum's funeral, I came back to the farm. Nothing had

improved there. The long milkings, not having the money to add value to the farm, crappy houses, plus I'd lost a manager because of injury — he'd crashed his car into a train. Everywhere I looked it was just awful. It was honestly the worst time of my life. Courtney was the one bright spark in my life at that time.

It was so rough that, for the first time ever, I really started to second-guess whether I even liked dairy farming and whether I wanted to stay in the industry. I felt a bit like I'd forgotten what my passion was about, but I'd signed a contract for two years and it still had another year and a bit to run.

Some of those doubts were settled when I became involved with a couple of organisations that were helping future farmers. I was still doing a bit of keynote speaking and around that time a couple of the Rotary clubs here in Invercargill wanted me to go and talk to them, so I did. After one of those speeches, Trevor Johnston came up to talk to me. He was the Chair of Venture Southland, an organisation promoting tourism, events and business in the region.

He invited me to a meeting with the mayor, Tim Shadbolt. I went in and had a bit of a chat with Tim and told him about the farming industry. I told him about the challenges that we'd had in the industry for a long time and about how they hadn't changed very much. As a result of that meeting with Tim and Trevor, I became involved with this movement called Southland Youth Futures.

The programme was run by Venture Southland, which is now known as Great South, and its goal is to help improve employment outcomes for young people in Southland. They organise for employers to do talks in different schools and help arrange workplace visits at places that interest students. They also align young people with their future employers then mentor them throughout that process. Pretty much, it does what Rob Sperling had done for me back when I left school.

Young farmers were getting jobs out on dairy farms then getting left to it. There really was no one there to monitor them or to have their back if anything went wrong. I saw being involved in Southland Youth Futures as a really good way of helping out some of those kids and making sure they stayed in the industry.

While I was getting into helping out Southland Youth Futures, I was asked to go on the Primary ITO Industry Partner Group committee for the dairy industry. These committees are made up of people from throughout the industry, including farmers, advisors and some of the big stakeholders like DairyNZ, and each one of the primary industries has one.

There are about 14 of us on the committee but only four or five of us are farmers. Our role is to make sure that all of the unit standards and criteria around being qualified in the industry align with what everyone on farms actually wants and needs. It's a chance for us farmers to tell the trainers what we needed when it came to our staff.

I've been on that committee since 2014, and what I've taken away from it over that time is that the problem in our industry isn't lack of knowledge or understanding, it's about the lack of delivery. The knowledge is all there if people want to find it, but once people have found it, sometimes the knowledge isn't delivered in a way that people can consume it and learn easily. In this new world, where people are able to get information very rapidly but can just as easily get side-tracked, we need next-level engagement. The content of a lot of the courses out there hasn't changed as rapidly as learning styles have changed, and we really need to do something about that.

WORKING YOUR BRAIN AND YOUR BODY

The problem with farming is that we're always physically busy but we can spend a lot of time being mentally bored. There's always heaps of jobs to think about, but once you've been doing them for a while, they all become second nature and you can do them with your eyes closed. That's when you run the risk of getting bored.

When your brain is a bit bored, you can end up spending a lot of time thinking about things you'd rather be doing but don't have time to do, or you can end up occupying it by worrying endlessly about any problems you might be having.

It's really important to either find things to keep your brain occupied or keep it relaxed while you're doing jobs that don't challenge you. For me, I find a huge synergy with fitness because the gym is one of the only places I go where I have to switch off. During a workout, my brain's occupied with what my body is doing and afterwards I'm so knackered that it just switches off. A solid workout is a really good way of shifting stress.

I can almost hear people out there saying they don't have time to go to the gym, so here's a workout that anyone can do.

FARM 4 LIFE WORKOUT OF THE DAY

DO 10 ROUNDS OF:
5 push-ups
10 sit-ups
15 squats

Once you've got these exercises down, start trying to do the 10 rounds as fast as possible. Make a note of your time each day, so you can see how much faster you're getting.

My best time for it is 15:36, so my challenge is to you is to try to get it down to under 15 minutes!

Back on the farm, we were both going to the CrossFit gym and Court was training at the weightlifting gym, but it was taking us each half an hour to get there and another half-hour to get home. I thought we could add a little bit more value and enjoy training a bit more if we had our own place. For the $400-plus a month we were paying to go to the gym — excluding petrol — I figured we could just buy some gear and work out at home.

Courtney was really getting good at lifting. She started talking about having a crack at getting into the Olympic team. Her coach was Jules Dempsey, who has twice coached the New Zealand weightlifting team at the Commonwealth Games, and he thought she had what it takes. It was fortunate that he worked for Ryal Bush Transport and coached at the Olympic weightlifting gym in town.

Soon, Court was training there twice a day — once before and once after work — but she was finding it pretty hard as she was always doing it by herself, and balancing it with her job was tough. We had a bit of a talk about it and I said I'd support her financially if she wanted to take time off work and give Olympic weightlifting a good nudge in the 63-kilogram category.

Courtney decided to take a year off work just as we were coming into calving in 2014. She helped out with feeding and looking after the calves, which was awesome as it meant I didn't have to pay a calf rearer. We had this little old four-stand woolshed on the place, and we had all our 980 calves in there after carting them back from the run-off. We rolled out five or six straw bales across all of the slats on the floor to stop the draught from coming through. The whole place just stunk of ammonia and Court and the other person we had helping both got cryptosporidium, which most of the calves had as well. You get crypto from ingesting a parasite that lives on poo, and it can make you really sick with diarrhoea, cramps and vomiting. It was awful.

Eventually, we had a couple of hundred heifers in one shed and 100 heifers up at another shed on the other side of the farm. Every day we had to cart milk from the milking shed to these other two sheds, which was a real mission. Everything was everywhere, which made it even harder for me to manage anything. It felt to me like everything was just as difficult as possible.

Just after we finished calving, Court headed up to Auckland to compete at the weightlifting nationals. In the run-up to that, she'd had to balance training on her own all the time and rearing the calves. Having crypto meant she couldn't train for the whole month before nationals because she was spewing all the time. That meant competing was way harder than it should have been for her, but she still managed to get fourth. Bloody awesome.

12.

PICKING MYSELF UP

Heading into the new year, things got even worse for me. Under a gentlemen's agreement, I was supposed to be getting 50 calves from the farm owner each year and he was supposed to take them right through to two-year-olds for me. I ended up rearing these calves and got them up to 18 months old. I'd organised to sell them to the Chinese market, who were paying big bucks for them — like $1800 a head.

The farm owner found out how much I was getting, then, all of a sudden, he decided he was going to charge me for all the grazing, back-paid over nearly two years. That stung me big-time. I sold the heifers, but by the time I'd paid for the milk, meal and grazing plus the tax on them, I didn't make anything when I should have been about $60,000 up on them. Add to that the fact I'd been spending way more on wages than I could afford as I'd based them on the milk income I was supposed to be getting, I ended up $140,000 in overdraft with nothing to show for it. I was working my arse off and going backwards.

I'd taken my eye off the ball in terms of finances and, because I didn't understand taxes well enough, I ended up in debt, too. Because I was so consumed with what was happening on the farm, I couldn't see some of the issues that were right in front of me. I had people in positions that didn't necessarily fit them, living in terrible accommodation, and I wasn't a very good boss. They weren't happy but I wasn't able to step back far enough to see that.

DON'T BE A DICK

Traditionally, farmers haven't wanted to drop their masks and show their vulnerability. That's slowly changing, but there are still a lot of people out there who work on the basis of 'that's just how things are done here'. That might have been how you were trained, but it won't motivate your workers to do their best for you. They're at a job because it's a job. They expect a warm house, a happy work environment and to get paid.

How you bring your values into your team should manifest in the months you have been working together before heading into calving. If you haven't had time to bring your values, goals and the way that you operate into your team before then, it's really important that you regroup often. Don't be shy to show vulnerability — sometimes as a boss if you open up to your team it can be awesome as it makes them feel like they can open up, too. You want them to think for themselves, think on their feet and buy into what you're doing rather than just being told what to do.

You want to make sure everyone on the farm knows why they're doing what they're doing, otherwise you're going to have to be the person telling everyone what to do. One way to address this is to call a meeting with the team. If you start having regular weekly meetings, you not only get to talk about what's happening on the farm and plan for the next week, but you also get to talk about the work morale, how everything's flowing, if you're sticking to the calving plan . . . all that sort of thing.

If you go three weeks without talking about that sort of stuff it can really build up and you'll end up with a big volcano of pressure that's about to explode. If anyone on the farm is welcome to call a meeting and it's not a negative thing, then they'll be more comfortable to have a talk about what's going on for them.

For young people in the industry who are working hard out and not getting the praise they deserve — I've been there and I remember the frustration of working hard and doing long hours but not getting the praise or the thanks at the end of the day. It's especially frustrating when your boss just doesn't see it. That's when you start to wonder what this is all for and if it's worth the effort that you're putting in. I can tell you now, it's worth it because your boss will see it whether they're telling you or not.

Even worse, I was in this relationship that I wanted to be in for the rest of my life, but I had to tell Court something that could have made her end it — I was more than $100,000 in debt.

When I told her, she just went straight into finance mode and her background really came into play. She put her knowledge into action trying to figure out how I could get out of this debt.

I was worried that I might not be able to get another job because it seemed to me that the guy I was working for pretty much hated me, my production per cow was pathetic, I was in debt — it was hard to see what I had going for me.

I didn't trust myself and didn't back myself. I don't know if I was depressed, but I was struggling to see the light because everything was so grey around me. I really started doubting myself. Doubt is like a huhu grub on a tree, it just eats away at you. I started to wonder if I should just chuck it all in, go bankrupt, head off overseas and go snowboarding in Canada or pig hunting in Australia with my mates. I didn't know what to do.

The first thing I did was ring the old man and said, 'Uncle, this is the situation . . . what do you reckon?'

He said, 'That was a bit dumb, wasn't it?'

'Yeah, it was a bit dumb, but what do I do?'

'Well, you don't go bankrupt!'

'What do you mean? That's the best thing I can do . . .'

'You can't do that. Look at all the people you're representing. Don't forget your last name and don't forget about us. You don't want to look like a failure. You've got to fight.'

At the time, I didn't want to fight, but I knew he was right.

I was in a really bad place. I just didn't know if I was good enough anymore. I'd won these competitions and felt like I had all this responsibility on my shoulders to be the up-and-coming young Māori farmer, but I just didn't know if I needed the pressure. I was in debt up to my eyeballs and I didn't know if I was even any good. All I could do was try, so that's what I did.

VULNERABILITY AND PERSPECTIVE

For me, showing vulnerability is actually showing strength. Whenever I see someone making themselves vulnerable in an open space where people can judge them, I see that as strength.

The weird thing is that when you're in that situation and you're showing that vulnerability, it can feel like you're showing weakness, so you try to hide it.

If you find yourself in that sort of situation, try to make yourself aware of the fact that what you think people are seeing and what they are actually seeing are probably different things. Your vulnerability is your strength.

Stepping back and changing your perspective is really important. On farms, it's something we can really struggle to do. A part of that is because you're always at work.

As farmers, we spend a lot of our time doing $20-an-hour jobs because they need to be done. It's really easy to get into the headspace of 'why the F am I doing this?' when you see all your mates going off fishing and diving and taking holidays. That can leave you feeling really negative about what you're doing. When you feel that way, it's important to remind yourself of why you went farming in the first place, to focus on the things that you do like and enjoy about your job.

It's kind of like when you go diving for kina. If all you're focused on are kina, then that's all you'll see and all you'll get. But if you lift your head up from time to time and have a look around, you'll see heaps of big fish, which will make you forget about the kina for a time. All it takes is a shift in focus and perspective for you to be able to get the kina, which is what you came for, and the added bonus of a nice big fish.

To relate that to farming, lots of farmers get so bogged down by all the jobs they have to do that they lose perspective on why they do what they do. When you're only milking, all you see is cows and milk. It's not until you go out to do some mowing that you realise how beautiful your farm is.

A lot of people get to the point where they wonder why they're even in the job. If that happens, it's really important to remind yourself why you started farming.

Is it because you love being outdoors? If it is, on those days when it's raining and you wonder why you're farming, try picturing yourself working at a job that means you have to be indoors. Not so bad now, eh?

Is it because you want to be your own boss? If it is, on those days when you're struggling to get the paperwork done and you wonder why you're farming, try picturing yourself working at a job where someone else dictates how you spend every minute of your day. Not so bad now, eh?

Trying to get my head around how to get out of both that situation and all that debt, I had a meeting with my accountant, Stew Perkins. He helped me to solidify my thinking about where I wanted to go. He also wrote me a game plan for getting out of debt, which helped me to start looking forward again. Having him believe in me was a massive hand-up.

Meanwhile, Courtney came up with a plan for our future. She was still working as an advisor assistant at Craigs Investment Partners, so she suggested writing up a combined CV for both of us that included all of her credentials as well as mine. She also explained the situation we were currently in and wrote about where we wanted to go.

Court applied for a heap of jobs. Having her write up the CV was brilliant because I'm bad at doing that sort of thing. I'm good once I get in front of people and can tell them why I should get the job, but getting the chance to get in front of them was all down to Court's work.

We sent the CV out to some of the farm advisors I really respected, and we had a phone call within 12 hours from one of them. Out of the six jobs we interviewed for, we only missed out on one of them.

We looked at the job offers and made a list of the things we wanted. We'd been living 45 minutes from town, and that wasn't good enough. I didn't want my darling driving an hour and a half every day so she could get to work. No one needs to waste nearly nine hours of their week just sitting in the car. Wanting to move closer to town took out two of the job offers.

We made the final cut for a contract milking job on a 500-cow property called Mirakanui just out of Woodlands, about 20 minutes out of Invercargill. The farm was owned by an investment company called MyFarm, which has a big portfolio of rural businesses, and was running a predominantly Friesian herd with a few Jerseys thrown into the mix.

A couple of the shareholders came down to check me out, so me and Courtney went out to meet them for an interview and to check out the farm. I think Courtney won them over more than anything!

When I got offered a job, if I could have used 20 hands to shake their hands on the deal I would have. I was so happy. It felt like a bright star in the sky for me. They took me on board and gave me the opportunity to run their farm despite what had happened on my previous farm. I'm pretty sure I got the worst reference ever from that farm owner, but that didn't change their minds. Wayne Taylor was one of the advisors and we clicked straight away. He ended up becoming a really big mentor of mine. I was lucky because he understood what had been happening on the farm I was on, and he could see it wasn't because of me. I was so stoked, and I am still so grateful for the owners and the farm consultant for trusting me and my experience despite the fact I was in a pretty bad place. Somehow, they could see a spark of passion in me that I couldn't really see myself at the time and took a big punt on us. What I realise now is that failing at that job taught me heaps of really important lessons, and I don't think I'd change anything about it — even though I'm still paying off the debt!

Wayne told us what his expectations of us were and what they wanted us to do with the farm.

One of the things that made me really want the job at Mirakanui was that I reported directly to the shareholders, who I knew were happy to take advice from their farm consultant. They had invested in a new effluent system and there was plenty of storage. On top of that, the place had pretty well-draining soils.

They were also planning on spending money on extending our buffer zone before the waterways, which involved moving our fences out another two or three metres, and there were plans in place to do a whole lot of riparian planting along the riverbed.

They were also in the process of establishing a closed system, which means that all the cows would be grazed on the farm. This lessens the environmental impact because they're not trucked off to graze further north.

I was excited to come to this farm and live in a house that was warm, with other good houses for the people working with me. I was also happy to be working for people who wanted to invest in their business.

Since I've been on the farm, we've reduced our cow numbers so that we've been able to keep them on the farm over winter. We went from a high stock number dairy farm that would winter everything off, to a low stock number dairy farm that wintered everything on. We bring in supplemental feed and also grow sugar beet, which is a high-calorie food crop, and kale for our winter feeding.

Another proactive thing the shareholders support us in doing is regularly blood-testing the herd, so we don't need to use as many drugs. Antibiotic resistance is growing, so we're careful about how we use these drugs. We herd-test four times a year, so we know which cows are likely to need treating and which ones don't. The ones that need a bit of extra support get drafted into a single herd, so we can keep an eye on them.

As well as riparian planting and protecting our waterways, we've spent a lot of money on technology. New technology allows us to manage the timing of the application of effluent disposal and to tightly monitor our fertiliser spread. We try to lead by example, and the farm's owners have invested heavily in getting the best advice around managing environmental issues. We all want to show the world what we do and that we're doing it to the best of our abilities.

MĀORITANGA

There's an old whakataukī that goes:

> He aha te mea nui o te ao
> He tangata, he tangata, he tangata
> What is the most important thing in the world?
> It is the people, it is the people, it is the people

That's pretty much how I look at the world — people come first. When you put people first, everything else follows. That said, I'd probably put the environment on an equal footing with people.

As Māori, we're all about the environment, so with all the new environmental rules and regulations around dairy farming that have come in, it hasn't changed anything for me.

I understand that a lot of people are going to really struggle with them — especially the new freshwater regulations that came into force in late 2020 — but for me it's not a matter of 'Why is it happening?', it's a matter of 'What do we need to do to ensure that we can continue doing what we love?'

Look after the environment and the people and you really can't go wrong.

———————

The year of 2015/16 was a low pay-out year, so everybody in dairy farming wanted to destock to try to reduce debt. When we signed the contract to the farm we were on, Court made a very good decision in favour of contract milking over sharemilking. If we'd gone sharemilking, we might have been bankrupt by now because of the pay-out going for a bit of a dive.

It was also the year I first put up a post on social media about farming, which turned out to be the birth of Farm 4 Life. At the time, there was a whole lot of media attention on people dragging bobby calves around, throwing them and kicking them. A lot of people seemed to think that that was how everyone treated their bobby calves, which was pretty tough for me to read.

I thought about it a lot while I was out mowing a paddock. When I'd finished, I saw three baby bunny rabbits sunbathing by a trough. I went over and picked them up, then set up my phone to film me. While I was sitting there patting these wee rabbits, I did a bit of a video about our process with bobby calves and how we treat them. I posted it on my personal Facebook, and it went viral. Heaps of people really liked it, which made me realise that I'd been able to add some value just by talking about what I do for work. I decided I was going to do more of it to help people understand farming.

Not long after that, one of my mentors asked me about my goals for the next three years. I said, 'I'm farming. I've got a plan to get out of debt over the next two to three years, we'll start to gain ground because I'm on a good contract at the moment. I'd like to start up a functional fitness gym with my partner, and I would like to design a video learning platform that people can go straight to for all things farming.'

I was still playing for Marist, so I decided to take my gym gear to the clubrooms and do some classes with the team as a bit of a trial for setting up our own place. There, we'd get 14 or 15 people turn up to do workouts with us, so I said to the boys, 'If I leased a venue and set up my own place, would you be interested in coming to support it financially?'

They all said they would because they really liked what we were doing. Setting up a gym was something I'd been thinking about doing for a while, but we really couldn't do it while we'd been living so far from town. Now that we were living only 25 minutes away, it was time to take the leap. That goal kicked in when we started up our gym called The Barracks in the old Army Drill Hall on Victoria Avenue in Invercargill.

The next goal to tackle was making a platform to help people learn about farming. I got thinking about how to build the platform from scratch and it all just seemed a bit hard at the time, so I parked that aspect of it for a while.

Then I had a yarn with Jamie McKenzie, who is a really good mate of mine from Marist. He was one of the coaches there and he's been a really good mentor of mine as well. I talked to him about this idea for having a hub. He reckoned it sounded like an awesome idea but asked me how I was planning to sell it.

Initially, I thought I'd do advertising in the newspapers and on the radio, but one of the bros, Tukes, said to me, 'Bro, you're pretty funny — that video of yours went viral so why don't you do a social media account?'

'You reckon? Would people watch it?' I wasn't sure.

'Yeah, I reckon! You're crack-up. You'll kill it. Do it.'

I decided I had nothing to lose, so I got into the whole social media thing in 2017. My goal was to get to 30,000 followers then I'd see where to go from there. I set up the Farm 4 Life Facebook page and posted my first video there on 12 September 2017. It showed me assisting a cow calving after it had been breach. Even though it's pretty wobbly and a bit all over the place, it's racked up about 13,000 views now. It took about a year to get up to 30,000 followers. It took a lot of work and a lot of editing for the videos, all while I was running the farm.

The earliest videos I posted on the Facebook page included ones about looking after your contractors, stabbing a cow with bloat — that one got me into a bit of trouble with the vegans — retention of foetal membranes, drafting springer cows and how to open calf meal bags without using a knife. I also included a few workouts for farmers and gave regular updates on my calves.

———————

By 2017, we were in our second year of running the gym and Courtney had gone back to work at Craigs Investment Partners.

Things were going pretty well, so me and Court took our first overseas trip together to Hawai'i. After we got back from there, Marist won the Galbraith Shield, which is the trophy for the top team in Southland premier club rugby. That was a big deal for me and the bros.

As well as playing for the club, we put a lot of work in around the place. I helped organise a whole lot of fundraising events including a couple of hāngī, a lip-synch night and a prize-giving ball to celebrate winning the prems.

Our trip to Hawai'i was followed by a holiday in Bali in 2018. Both of those trips made us want to do some more travelling, so we started talking about doing a bit of an OE. It was the perfect time to do it as I had good staff working for me and really good farm owners.

We planned to go to Europe for seven weeks in 2019, but first we had to make sure the owners trusted the people we were going to leave running the place while we were gone. We still had eight months to put everything in place to make sure everything would run smoothly while we were away.

To do that, we worked out we needed to get the right people working with me on the farm. When I'd arrived at Mirakanui, I employed an older guy Aaron and a couple, Hinekura and Michelle. Hinekura was the farm hand while Michelle did the calf rearing. We all worked together for quite a wee while, but eventually Michelle and Hinekura split up. When one of them took off up north, the other followed, leaving me and Aaron to run the farm for a bit.

Aaron moved on at the end of the season, which was when Paul and his son DJ came to work with me. Paul was a really good worker, while DJ was just learning the ropes.

That taught me a lot about how to manage staff who were also family. There was a real challenge for me because they spoke to each other as father and son, and not as co-workers should speak. It was a real learning curve for me.

DJ and Paul had a bit of an argument, and DJ left. Then Paul got injured towards the end of the 2017/18 season, so he was on ACC and still living in the house on the farm.

It got a bit complicated as it meant I couldn't fill his role because I couldn't offer accommodation for the time being, so I was running the farm pretty much on my own. The boys at the gym knew I was in a bit of strife, so one of them, who was known by everyone as Banana, told me that his cousin from Tokoroa had just come down to Invercargill and he wanted to have a crack at farming.

Josaiah was 16 and a really solid kid. He didn't really know how to drive a car and was missing a few other basic skills, but he was keen to come and work with me in the school holidays. I was happy to have an extra pair of hands for calving.

He wasn't liking school much and he really enjoyed being out on the farm working. In the end, a few other boys from Southland Boys' High School came out with him to help out chipping weeds and all that sort of stuff.

After the holidays, Josaiah didn't want to go home, so he ended up staying for a little bit longer than we first expected. He did bloody awesome for a 16-year-old. I was really happy to be able to give a young kid a chance. He had all the potential in the world and was at a stage where he was wanting to learn, and once he was with the right people he picked up things really quickly.

Eventually, he decided to go back home to Tokoroa to spend some time with his family and catch up with his mates. Before he left, we threw a big birthday party for him. We had a big bonfire, and we put up a tent, took a generator down and brought all the sounds down there. It was a bloody good night and everyone enjoyed having a bit of a hoo-ha on the farm.

Not long after Josaiah left, my mate Tuki arrived back from the UK and agreed to come to the farm to help out. He's helped me out quite a bit over the past four years and I'm always happy to have him when he wants work.

After that, I had a couple of young guys work for me who didn't last long. They both started off with big plans then, within weeks, they both ended up heading up country to chase their partners.

After four weeks, one of the guys just didn't turn up for work one day. I got down to the shed to take the second mob in for milking and I was super impressed with what a good job he'd done cleaning the place down after he'd milked the first mob. It took a few minutes before I realised the reason the place was so clean was because he hadn't milked them at all.

I brought the cows in and milked them, then tried to get a hold of him. He wasn't answering his phone, but I gave him a couple of days. I thought something must have happened and he'd come back eventually.

After two days went past, I went down to the house and found he'd abandoned it. He just packed his stuff in his truck and disappeared.

As a result, in June 2018, I was on the farm by myself with calving coming up. It pushed my limits a bit as there was just me and I was pretty much two staff down. I decided to chuck a message up on my social media to see if there were any farmers out there who could help a young fella out who was in a tight spot with staff.

The power of social media helped me claw myself out of the hole and I got a phone call from a beautiful lady called Kerry. She was a sharemilker, but she was between jobs. She and her husband came out and helped me with milking, then they stayed on and helped me with calving, feeding out and generally helping me around the farm. Their son even came out and did some tractor driving and cattle drafting. As well as being an awesome worker, Kerry was also a mean baker. It was wicked getting to the cowshed and finding some sneaky snacks that Kerry had cooked up for me.

I really appreciated everything they did as they helped me out of one of the worst positions that I'd ever been in. They did a great job and having them there gave me time to employ awesome staff.

While Kerry was still there, I put an ad in the *Southland Times* looking for a 2IC/manager. That was when I was approached by Guri Sohi. I told him to meet me at the shed. He turned up and he seemed like such a polite young man who really wanted to do well in the industry.

After I offered him the job, he really cemented his value for me when he said he couldn't start when I wanted him to because he couldn't leave until his current employer was in a position to let him go. As soon as he said that I thought 'Yes, he's the one for me!'

Guri is from Punjab state in Northern India. He had friends in New Zealand, so he decided to come here to study IT in Christchurch in 2014. After he finished his course, he got a one-year work visa and couldn't get a tech job. One of his mates worked on a dairy farm and used to tell him stories about farm life, and that's how he ended up working in farming.

When I first met him, his wife was still living in India with his parents. Guri's dad's retired now, but he was a farmer in India, but he wasn't a dairy farmer because cows are sacred over there. Instead, they milk buffaloes.

A little bit after Guri started, Kerry got a full-time job. It was time to put my feelers out up north again to see if there were any youngsters up there who wanted to come down and work for me. One of my cousins up there told me his stepson, Detroit Ririnui, was keen.

Detroit was 20 and he's from Matakana Island. He'd just come back from Aussie where he'd been working as a barista. He was bumming around doing not much up in the Bay of Plenty, so I offered him a job. He came down and he's gone from strength to strength ever since.

He and Guri did an amazing job sussing out how everything worked and learning about how things were done. If me and Court were to have any hope of going on our dream OE, I knew I needed to show the farm owners that we had a strong manager and a good 2IC on the place. As soon as I could, I began including Guri and Detroit in the meetings with the shareholders and the consultants, which helped everyone get comfortable with the idea of us being away. We were so lucky to have such supportive owners that we were allowed to go away for that length of time.

One other thing we did to get their support was to make sure that they knew we had a really good relationship with all our neighbours. Ringo and Nicole live on the property on the south side of the farm. They're about the same age as us and we talk often. We know we've got each other's backs if anything happens.

On the western side is Andrew. He had the same farm consultant as I did for quite a while, so we were good. We used to have a bit of a production competition because we had the same person advising us! Ross is the sheep farmer on the property behind us. He's the guy I get mutton off if I'm after a two-tooth or a lamb. Me and him will do a bit of a trade. If he needs colostrum milk for his calves, he'll come down and get it from my place.

With such great people living around me, I knew that if the shit hit the fan and things went wrong while we were away, they'd all help out. They all knew we were planning to go away, and they agreed to drop in once or twice a week while we were gone, not to check up but more to see if the boys needed anything.

MONEY TROUBLES

Everyone wants money, and sometimes it takes borrowing money to create money, but when the creation side of it doesn't happen, people can find themselves in a bit of a hole. I went contract milking, invested in animals, went into a dodgy gentlemen's agreement, and was very naïve about my tax obligations. I got burnt and ended up in a financial hole. I realised I was in the hole, got some support around me, made a plan and am getting out of it. The whole thing was a huge learning curve and it's taught me some good lessons.

There's no simple way of getting out of this sort of situation, but it's important to talk to people as soon as you realise things aren't going to plan. That way, you can get in early and put things in place to stop things from becoming worse. It's not going to get better if you ignore it, or you just hope it's going to get better. Get in and make changes before it all blows up in your face.

Sometimes you have to just lay down your pride and step over it to get help. Taking a step backwards might feel stink, but no one will think any differently about you for doing it. In a pure business sense, it would have been far better for me to go bankrupt and just start again. But I didn't do that because I didn't want to be known as that guy who went bankrupt. Instead, I just dug out of it.

If you're avoiding answering the phone when someone from the bank rings, remember that the banks don't want you to go bust — they want you to pay their money back — so they'll help you as much as they can to find answers and help you out. First, though, you need to be honest with them about what's going on. It might hurt your pride a little bit to face up to and talk about it, but it will help you heaps in the long term.

I've been there — avoiding the bank, avoiding my accountant, avoiding the IRD — and on the twentieth of the month when I got my pay-out, BOOM!, it was all gone. I had three families to support and the stress of not being able to pay them was unbearable. Eventually I realised that I had to draw a line in the sand and make those calls that I'd been avoiding for three weeks. On each of those calls, I arranged to meet with the people in person. At those meetings, I told them that I'd stuffed up big time, explained my situation and asked for help.

Every single one of those people helped me to create a four-year plan to get financially stable again. I know now that if I'd done that three months earlier and had a three-month headstart on that plan, it would have been a totally different story and things would have been much easier.

No matter how tired, anxious and stressed you are, no matter how hard it feels, if you make those calls, you'll find people who want to help you to make things better.

13.

UNDER WATER

Apart from my darling Courts, there's one other big thing that has kept me down south. I've done a lot of stuff — I've done a bit of snowboarding, I like skiing, I've played a lot of rugby and done a lot of pig hunting, but diving is definitely my best hobby.

Uncle Kelly is a pretty good diver. He used to dive all around Australia when he lived over there. He showed me some photos of really big crayfish he used to catch. Because Uncle was into diving, I got into it, too. I still have a pair of fins he used to use. He's in his early seventies now, so they're pretty old. I love having them.

When you go diving, the opportunity to get out there and get everything right is very rare. You can go pig hunting in the rain, you can go pig hunting in the snow, you can go pig hunting in windy-as weather. You might not get anything, but you can still go and look. But with diving, everything has to line up before you can even get out there.

Down here in Southland, we need four or five days of a northerly wind blowing, we need to be 10 to 15 days out from any storms, and we need to time the tides right. On top of that, it can be hard to find a time that lines up for me and my mates. There are very few divers I'll dive with. It's pretty rare to find a friend who's a good diver and has the same availability to go diving as I do!

Trying to line up all those ducks and then fit in a dive around work can be really difficult. That's why when I decide to go out, I just go. There's no hanging around waiting to hear back from anyone, I just get out there. A lot of my friends will be like, 'Oh, you could have given me some notice!' There isn't time for that. You've got to be ready to go when the time is right.

I'm lucky that I have two good mates who have the same approach as me, who I love and trust, and who are into the same things that I'm into, so we go diving together quite a bit — Dan, my mate from Marist, and Rory.

A few years ago, Dan bought a really cool boat, which gave us both a good excuse to head out diving whenever we can. Dan's a lot like me in that he's always ready to go for a dive whenever the conditions are right. It's pretty typical for one of us to message the other at about three in the afternoon: 'Feel like a dive?'

'What time?'

'5 pm?'

'Sweet, see you at Bluff.'

Two hours later, we're there and in the water, even though Dan lives near Tokanui, which is on the way to the Catlins. It takes him about an hour to get to Bluff, but that never puts him off.

He's always keen to go when conditions are good. We don't go out without each other very often.

From Bluff, we'll shoot out to dive around Ruapuke Island or some of the smaller islands off it like Green Island and Bird Island. Ruapuke is about 15 kilometres southeast of Bluff and 30 kilometres north of Stewart Island. The island itself is about 16 square kilometres and it's covered in scrubby bush. The waters around it are beautiful.

I met Rory when he started coming to the gym. He told me he was a fisherman and asked if I wanted to go out on an oyster boat one day. I was keen as, so Rory teed it all up. It was unreal. I didn't realise how they operated the dredges, though.

When they bring the dredge up to the side of the boat, they have to rock it to get it to release all the oysters. Even though I'm used to being out there in a boat, that rocking combined with the waves out in Foveaux Strait meant I was as sick as a dog. I still managed to do my mahi for the day, but I was sorting oysters then leaning over to have a good spew. It was an awesome experience, though. Those guys who do that for a job are wicked. And, by the time I got home, I'd recovered enough to have a good feed of Bluffies, which are the best oysters in the world.

Not long after I'd been out on the boat with Rory, he said, 'Do you want to try for a dive?'

'Yeah, I'll come for a dive.'

He said he was taking some guys out for a fish and dropping one of them out on the Tītī Islands, just south of Stewart Island.

I was like, 'Can you dive?'

Rory was shocked. 'What do you mean, "Can you dive?" Of course I can dive! Why would you even ask me that?'

I told him, 'Most of the people who ask me if I want to go diving with them only want me to come so I can get them a feed. They can't even dive.'

He was absolutely shocked. He's a really good diver and he knows a lot of good spots down around the islands in Foveaux Strait.

Since then, we've dived a lot together, and he's helped me out heaps on the farm. When he gets sick of fishing, he'll park the boat up and come and do some building or some tractor driving for me.

When we go diving anywhere, Rory's like the patriarch of our dive group. Being a fisherman and coming from a fishing family, he knows a lot and we trust him. He's a very technical diver. He plans his dives and his descents really tightly.

The reason I like diving is the fear factor. In the water, humans are not the predator, whereas on land we can take on anything. When you go diving, you can flounder around on the rocks close to shore and get nothing, or you can take a risk and swim out to those rocks that no one swims out to (for obvious reasons) and dive out there.

WHAT DIVING HAS TAUGHT ME ABOUT LIFE

Diving is one of the only places in New Zealand where humans aren't the apex predator. If you're out pig hunting or deer stalking, you've got a gun so you know you can take on whatever is coming your way. If you're out in the water and you see a great white shark, shooting it is going to do nothing. It will just think, 'That was a cool toothpick, but now you've pissed me off!'

Recognising your place in the order of things and understanding this makes it easier to get comfortable with being in the water. Diving also makes you realise the importance of being cool under pressure and it helps you to enjoy working under pressure. Fear is something that is always present when you're diving, but you have to try to relax and not let it affect you. The adrenaline might be pumping, but you have to try to calm it in the face of the unknown.

A big thing with breath holds for me is that when you feel like you want to gasp, you're only halfway and that's when you really start diving. I relate that back to business. Sometimes, you want to pull out because you feel like you're not in control of what's going on or you're worried about the level of risk you're taking. Everyone feels that but the majority of people pull out rather than carrying on. That moment is the difference between getting the big crayfish or no crayfish.

For years, I've been diving down here and it wasn't until I went from diving in six metres to diving in eight metres that I went from never catching crayfish to catching four or five in one breath — it was all because I dived a little bit deeper instead of giving up. Being able to switch off that urge to pull out and to trust your body and your gut means you can get the bigger rewards. You've got the choice — either chuck in the towel or trust yourself.

It's scary. Every time I go down, I'm pushing my body to the limit. I never do a descent and come up with oxygen. I always come up out of breath. I love the challenge of it.

When I'm diving, I can zone out from everything that's happening around me. All of my senses turn on. It's like I know I'm in the danger zone — it's a bit like being in the middle of a fire and not knowing where the exit is.

It's also very tactical. Fish don't just swim up to you and let you shoot them. Depending on what species they are, they'll behave differently. It's hard enough to get out there, but being able to get out and catch a feed makes it worthwhile.

We get blue cod, trumpeter, blue moki, butterfish or greenbone, and sometimes we'll see some little tarakihi. There are kingfish down there as well on the odd occasion. I've never seen one yet when I've been spearfishing, but other people I know have got hāpuka. Then we also dive for crayfish, big pāua, mussels, kina, scallops, oysters and the odd octopus.

We might go out every couple of weeks and dive on the same rock, but I'll always have a different story to tell about that rock every time I dive on it. There might be four crayfish under it and I'll take all four of them, and in a week's time, there'll be different crays in those holes.

It's targeted killing. When we go down, we look at all the fish species down there and make choices about what we want to take. I always pick the type of fish I want to cook. I'll take a breath and go down, then I'll see a trumpeter, a blue moki, a blue cod or a greenbone. I'll think, 'Hmmm, what are we going to have for dinner tonight?'

Are we going to have homemade fish and chips, because if we

are, I'll get some blue cod. If we're going to a barbecue somewhere, I'll take a trumpeter and smoke it to take with us. Trumpeter are a bit rarer to find, but I really rate them. Everyone raves about how delicious blue cod is, but I reckon trumpeter trumps them every time.

If I want to drop a nice fish off that's going to feed a big family, then I'll get a blue moki. They're not the best-tasting fish, but they're a really big fish with plenty of kai on them. They're also quite dumb, which makes them easy to catch. I've had times when I've speared them and my spear has deflected off their scales, but they've turned around to come and have another look at me!

I love diving as it's a good way to get kai to give to people, but Uncle always taught me that you shouldn't freeze kai moana. He'd say, 'If you want a crayfish, go out for a dive. You never want to be that guy who says "I want a crayfish" then goes out to his freezer and gets one from six weeks ago.' His advice was always if you can't eat it fresh, give it away otherwise it's a waste.

That was so ingrained in me from a young age that I just never think of putting seafood in the freezer. If I'm getting crayfish and I'm getting four or five of them, we can only eat so much so I give the rest away. People think you're bloody Santa Claus when you turn up at their house with a cray for them.

I was just a duck diver up until about 2018. Once I started diving with Rory more, he showed me a different way to approach things. He has really cool, calm, relaxed mannerisms around diving. He started telling me about how relaxing a little bit would slow my heartbeat and prevent me from going into fight or flight mode when adrenaline gets released in my brain. When adrenaline gets released, it speeds up your heart and your breathing, so you use up your oxygen much faster. When you're free diving, that means you can't stay underwater as long between breaths.

Rory decided to teach me a bit more about holding my breath for longer. We went to Splash Palace in town and took our dive gear in there. We did a max breath hold, then we spent a couple of hours in the pool with him teaching me these breath-holding techniques.

After a few hours, I'd tripled my breath hold. Being in the pool is very different from being out in the sea as there's nothing to look at except the black stripe on the bottom and there's nothing in there that wants to kill you! Even so, it was a real confidence-builder for me to realise that if I could control the thought patterns that go on inside my brain, I could extend my breath hold significantly.

He taught me a different style of diving, and he also taught me that he doesn't worry about sharks or anything because he knows he's either going to live or die and thinking about it is not going to change the outcome. It's not like you can prepare for it if you see one.

Rory helped me to have a different perspective about the risk of running into sharks. He has such a good perspective on it and he handles it really well. I always try to put it right to the back of my mind, but it usually stays near the front. It's right there.

One day, we decided to try to find some new possies to dive down the back of Bluff, just off where the old Ocean Beach freezing works are. Finding new spots to dive in is always a bit of an adventure because you're never quite sure what you're going to encounter.

We were about 200 metres offshore when we saw this big rock

formation under the water, so we decided to go and have a closer look at it to see what was going on down there. We hopped off the boat and swam over to the formation, which turned out to be a pinnacle that was about eight metres high but completely submerged. It was awesome.

One of the things that make Rory and me good dive companions is that we're both quite strict on safety. We nearly always dive one down, one up, so there's always one of us at the surface keeping watch for the other one.

Rory went down and, because the water was quite hazy, he soon disappeared. I started to breathe up, which involves taking deep breaths for about a minute, then, in the last 10 seconds before descending, taking three quick breaths followed by one really deep one before going down.

I could just see Rory coming back up when I was ready to head down. I started descending when he was about two metres from the surface. As I went down, Rory came shooting up beside me heading for the surface. He seemed to be going a bit faster than usual, but I didn't really think anything of it.

Because I'd been preparing for quite a deep dive, I was quite relaxed, so I didn't really take much notice of the grey figure I could see below me. That was until I realised the grey thing was a big sevengiller. Sevengill sharks can grow to a couple of metres and they're pretty opportunistic when it comes to feeding. They're one of the most common sharks seen around New Zealand's coastline and they are quite well known for being aggressive towards divers.

As soon as I realised what it was, I decided to swim straight back to the surface. It turned out Rory had been heading for the surface more quickly than usual because the shark had been chasing him! We were quite a way away from the boat, so the pair of us were back to back on the surface of the water with our spearguns ready to go when the sevengiller swung past us, smashing me in the back of the leg with its head as it went.

Given that the shark had chased Rory and whacked me, we knew it wasn't going to leave us alone any time soon. It was pretty relentless about getting us out of its territory. It swam around and came back again, so Rory stabbed it in the head with his spear and I swam down after it to try to chase it away. I managed to stab it quite hard with my spear again and it took off. I was pretty confident it was gone, but I'd been so worried about the shark that I didn't equalise my breathing and my ears started ringing big time.

I shot back up to the surface to take a breath and got back to Rory. Just as I said 'I think it's gone, bro!' the bloody thing came back!

It was right underneath our feet, but thankfully by then we'd managed to make our way over towards the boat. We both swam for it and jumped back on board, happy to be out of the way of the apex predator in the area.

After that, we went diving way out the back of Ruapuke Island in a place called 'Spot X'. It's like Jurassic Park underwater there. There are these huge boulders that are about eight metres tall. These big rocks are all piled on top of each other with heaps of nooks and crannies to explore.

One of them, we swam down about six metres then swam under the rock about two metres, then down and in another couple of metres. It opened up into this big cavern. In there were all these big crayfish, and there were big trumpeter and blue moki swimming about. The water was so clean and clear it was unreal. It's amazing.

Every breath, every dive, I wonder what I am going to see. The first time we went there I couldn't help but wonder if anyone had ever dived there before. I'd swim into corners in the rocks and I knew I was probably the first person ever to have seen that place. It was absolutely amazing. The buzz of it was unreal.

From diving down there, we started planning trips into the Sounds in Fiordland. It's so beautiful and untouched in there. It's truly magical. Courtney's pōua used to talk about all the caves that are up in the Sounds. Heaps of chiefs have their bones up in those caves. There's so much dense bush up there you'd never find them, but they're there and their wairua lives on in the place.

It's such an amazing trip that we'll usually have guys flying down from Tauranga or Canterbury to come with us. We've even had one of the guys get helicoptered in and dropped off on the island — that was some superhero kind of shit!

When a bunch of us decide to head into the Sounds, we'll give the Department of Conservation (DOC) a call and book into one of their huts there. We usually stay at either Gut Hut, on Secretary Island, which sits between the mouths of Doubtful and Thompson sounds, or Deas Cove Hut, which is just across the water on the shores of Thompson Sound.

Gut Hut is pretty basic. It sleeps six people and has an open fireplace for heating. Over at Deas Cove Hut, things are a bit more flash. The old hut that was there almost got wiped out in an avalanche, so DOC replaced it in 2008. The new one sleeps 10 and has a wood stove. Of course, we always want to go to Deas Cove if we can!

Dan will bring his boat in, and if another one of the boys wants to bring theirs then we'll take a couple of boats and maybe a little dinghy so we can zoom around. As well as the boats, we'll take about 100 litres of fuel, about six days' worth of food and a cooker.

We leave Invercargill at about 4.30 in the morning and head up to Manapōuri. There, we launch the boats into the lake and head over to West Arm, where the Manapōuri Power Station is.

On one trip, I took Uncle Kelly with me. He never imagined he'd be going in there and he absolutely loved it. Even more awesome was that his brother had worked on the power station over there in the late 1960s. Uncle wanted to go and have a look around the power station. Where they used to have all the machinery and stuff is all covered in bush now, but there used to be a big clearing there. It's amazing how quickly it's gone back to being covered in bush. I introduced the old man to all the sandflies down there, which he didn't love so much.

From the power station, there's some guys who chuck the boats on trailers behind their trucks and tow them over the Wilmot Pass. They were there waiting for us and the boats. These guys know the road really well and I was happy not to be driving over it. The road across the pass is pretty hairy. It was built back in the 1960s so the workers building the hydroelectricity project could get across from West Arm to Doubtful Sound, where the lake water used to generate electricity at the power station flowed out to sea. It's 21 kilometres of windy gravel road that climbs over a 670-metre mountain pass.

When we got down to Deep Cove at the head of Doubtful Sound, we launched the boats and headed on up to Secretary Island. We stayed at Gut Hut and it was amazing. We took a few tents in there with us because there were more of us than the hut could handle. When we pulled up outside the hut, we all rushed off the boat. Dan was safe because the boat owner always gets a bed. The rest of us had to fight for the remaining five bunks. The first ones to get their bag on a bed were sleeping inside. The rest of us had to either sleep on the floor or in a tent.

Once we had our camp set up, we went out for a dive. We got three big hāpuka and four little schoolies. We cooked them up for a couple of dinners. The next day, we went around the other side of Secretary Island and got some of the fattest mussels I've ever seen in my life. They were big, big, green-lipped mussels. They were just so fat and sweet.

Not far from where we got the mussels, we were catching good-size buck crayfish in waist-deep water. I didn't even have my mask on. I literally just put my hand down, reached under this rock and grabbed a big crayfish. It was unreal.

From there, we went up to one of the other bays where we found heaps of horse mussels or kūkukuroa. They were huge, man. They were some of the biggest horse mussels I've ever seen. We got a big feed of those as well. We got heaps of kina and pāua as well.

Over there, we saw a couple of big sharks — they were seven-gillers. They're not really known to attack people, but we kept out of their way anyway. Then we saw a few kingfish down on one of these ledges. A couple of the boys, who were diving on bottles, headed down to see if they could get them. I thought, 'What better time to see how deep I can dive?'

I dived down through the boys' bubbles and got down to where they were. I gave them a bit of a fright because they did not expect anyone else to come down there. I had a bit of a look around, then they gave me their crays to take back up and I grabbed an extra couple for myself while I was there.

Cooking the food on the fire, everyone eating and drinking together with candles going, and the stories that you tell when you're around the fire, it's such a cool experience to share with your mates. There's no cellphone reception and no interruptions, so everyone is present and in the moment.

On the way back out from Doubtful Sound, we always start planning our next trip. We've decided we want to do it every year. That's a great thing to do as it means we've always got the next trip to look forward to. Then when we get out of there, everyone is geeing for a shower, a nice bed and some clean clothes that aren't damp!

I've dived all over the country. I've been up to Doubtful Sound and Milford Sound. I've also done trips out around Kaikōura, Kaiteriteri and the Abel Tasman National Park. Up north, I've dived all around White Island and Mōtītī near Tauranga. I've been in the water all around Tolaga Bay and the Coromandel. The only place I haven't dived yet is up in Northland, so that's on the list of places to go one day.

I was diving up at White Island only a couple of weeks before it erupted in 2019. The visibility there was unreal. When the water's really clear, it feels like you're flying under the water. I saw this snapper right down the bottom. I swam down to it, shot it pretty much right under the boat, then pulled it up with me.

The guy on the boat was like, 'Oh, did you just get that fish on the bottom?'

I said yeah and he looked a bit surprised.

'That was twenty-two metres, bro!'

That was one of the deepest dives I've ever done. But, when I can see everything really clearly in the water and I dive down, it's like I lose all of that fear because I feel so comfortable. When I go out and I can only see for one or two metres, my senses will be lighting up all over the place and the connections in my brain will be going nuts.

We were just about to leave the island, and we'd seen a couple of kingfish floating around here and there. I swam down really deep to where the water was a deep, dark blue. I wasn't expecting that much, and I was a little bit scared. When I looked up, I saw what looked like a train going past me. It was a big, thick mass of kingfish. It was crazy!

There were so many of them and they were that thick that I couldn't even think about shooting them. I had my speargun, it was loaded up, and I just sat there and admired them. It was absolutely amazing — just such a beautiful thing. It's never all about getting fish for kai — it's more about the experiences I have when I'm diving and being able to take that time out away from the farm and the gym in a space where it's just me, my mates and the sea.

TAKING A LEAP OF FAITH

I know first-hand that when you go diving the big kina and the big crayfish are way out in the depths of the uncharted waters. You don't get to see those ones without taking a decent level of risk, having a good amount of self-doubt and an equal amount of heart. When you're out there trying to catch those fellas, right at the front of your mind will be any possible threats or looming shadows.

It's exactly the same with any kind of business — whether you're going from 2IC to manager, or manager to contract milking — that self-doubt will always be there.

If you have people saying you're doing an exceptional job with what you're doing right now and you should move up, that's because they've seen you have capacity to add more value to the business you're currently in or to the business as a whole. They aren't telling you that because they think you're going to do an exceptional job straight away in the new job, but because you have the potential to do it.

It might mean moving from where you are now to another farm to move up the ranks. Obviously, that comes with more money, but more money means more responsibility.

It does get boring if you're always just floundering around the rocks and wondering what's out there further in the deep water. Just remember you can always come back into the safety of the shallow waters if you sense danger.

In terms of business, that means if you go from a manager's role to contract milking and find you don't like it, you can always come back to that manager's role. Once you step forward, you can always take a step back, reassess then move forward again.

Don't be scared of that. People are scared about failure too much — just jump in, take that leap of faith. There are heaps of people around you to support you. All the businesses that support the industry are happy to share their knowledge because, unlike any other industry, we all get paid the same for our milk.

T'S BEST KINA RECIPE

Get some kina (preferably dive for them yourself).

Stick them in the shade for three days.

Crack them open.

Eat them.

Delicious!

14.

A NEW LIFE

That eight months went by real quick and soon we were packing to go to Europe. I was pretty excited because I'd sacrificed doing my OE when I was younger in order to establish myself financially. I don't regret not travelling when I was younger because the sacrifices I made then have got me to where I am now. I've done all these things so I'm in a position today where I can run a multi-million-dollar farm and then travel around the world for three weeks every year.

Getting time off when you're self-employed can be a bit tough. Sometimes I find there's so much organising that goes on before I go on holiday, it hardly feels worth it just to have a few days away. The stress of that organising can be so massive that I feel like I might as well stay here and do the work myself! That's a big trap.

When you don't go away because it's too hard to find a relief milker, or you can't afford to pay someone, or whatever, you not only miss out on having downtime away from the farm, but you lose that chance to step away from the farm and get some perspective on how things are going within the business and seeing all the things that you can work on.

As a farmer, when you travel in New Zealand and you get a business phone call while you're on holiday, not many of us have the balls to say, 'I'm out fishing' or 'I'm on holiday, I'll get back to you when I'm home'. Even though it can be hard, I reckon it's really important to say no sometimes — espcially when you're on holiday.

Travelling overseas was bloody cool because I didn't have my phone switched on and I could only be contacted by email or through Facebook — and I could choose when I logged onto those things. Although, while we were away, I was still having to do regular reports for our farm consultant and our shareholders, so I still had my fingers on the pulse of what was happening back home.

After a whole bunch of flights from Invercargill to Christchurch then to Sydney and on to Dubai, we finally landed in Dublin. We were in Ireland to meet Court's cousins there. I documented the whole trip for the Farm 4 Life page.

After a visit to Blarney Castle in Cork, I went to meet a dairy farmer, Paidi Kelly near Kilmallock, just south of Limerick. Then I headed to Teagasc in Cork, which is where the Moorepark Dairy Research Centre is. There I met Bill O'Keeffe. He talked to me about the work they were doing studying grass, cows, fertility, calf rearing — you name it, they were researching it. The dairy industry in Ireland is growing really fast, so it was good to talk to the guys there about how things are changing and the opportunities that are coming with that change.

They were awesome as they shared heaps of knowledge with me and I learned a lot about how their systems operate and what their challenges are, and I got to share a bit of knowledge about how we run our farms.

In Ireland, I caught my first train ever back to Dublin, where at the world-famous Temple Bar I met up with a couple of old rugby mates who had played for Marist. It's fair to say we smashed back a few pints of the good old Guinness before the boys took us on a drunken history tour of their city.

From Ireland we went to London, where we caught up with a whole bunch of our mates from home. We stayed with the Sim twins, Nikki and Emma, who are from Invercargill and are both working as lawyers over there. Our big night out in Dublin was followed up with an equally big night out in Brixton. Far out, it was awesome.

Exploring London was amazing. The history and how old the country is just blew my mind. When we come from such a young country, it's hard to believe there are buildings that are like 1000 years old that are still being used. The other thing I found hard to get used to was the huge number of people everywhere.

From London, we joined a Contiki tour for 28 days. During that time, we visited 10 different countries. We caught the ferry from Calais to Paris. There we ended up accidentally getting amongst a big protest that was storming up on the Champs-Élysées. It turned out to be a bunch of people in yellow jackets protesting about high taxes and the price of petrol. They ended up smashing a whole lot of shops, so the tear gas and water cannons came out. That was pretty loose. Even looser was our night at the Moulin Rouge. I hadn't seen that many boobs since I left the farm!

After that little bit of noise, we headed for Switzerland. We happened to be in Lucerne when they were having their annual celebration of the end of winter. The whole town was on lockdown and everyone was getting dressed up to go out and celebrate. The festival is called Fasnacht and everyone wears ugly masks to try to scare away the winter.

I said to Courtney, 'Come on then! Let's go and get big-time dressed up!' We purchased these really flash outfits, but it was so cool. It really made us feel like we were part of it all.

Everything was closed and everyone was out on the piss. They were all wearing these incredible costumes that looked like they belonged on a stage. They must have spent thousands of dollars to dress up for that one night.

They were carting around these trollies with warm alcohol in them, and there was this awesome melted cheese called raclette. There were probably 20 different brass bands, each with about 20 or 30 people in it in full costume. They played the coolest music — like Limp Bizkit, Taylor Swift and the theme from *Star Wars*. I was so glad that we made the effort to get into the spirit of the whole day. It was the most incredible experience, which was even better because we hadn't planned any of it.

From Lucerne, we bussed on down to Lyon, then to Barcelona. From there we headed back to France and along the Riviera to Monaco. We crossed into Italy, where we visited Florence, Rome, the Vatican and Venice.

My favourite thing in Rome was seeing the Coliseum. I'd

wanted to see it since I was 13 when we learned about it in social studies at Tauranga Boys' College. I was so into it back then I even built a model of it, so it was unreal to actually see the real thing.

In Venice, there was a coffee shop that had been there since 1720. That's older than our country! New Zealand wasn't even New Zealand when they first started selling espressos ... unreal.

From Venice, we headed to Austria, then to Germany and on to the Netherlands. We ended our tour in Amsterdam while the rest of the crew were driving through Belgium on the way back to London.

It was weird not being with the mates we'd made on the bus. We felt really alone having always had company for the past month. We were pretty happy to get on the plane home from Amsterdam. We had a couple of days in Dubai on the way. It's a weird place. It has no character or personality. Everything just felt a bit fake.

I was really happy when we finally arrived home at the end of April, ready to get back into the busy job of drying off all our cows. While we were away, our coaches at the gym had done an awesome job, and the boys on the farm had kept things ticking over. We made sure to take some time to celebrate the wins of all our staff. We were so fortunate to have these people around us who were so supportive of us slipping away for seven weeks.

While we were away, Farm 4 Life grew to 100,000 followers because I'd been uploading stuff over the whole trip. That was my real call to action to start thinking about what my dream of an online hub would look like and what I was going to do in order to create it.

That trip to Europe gave me a new perspective on things. It

made me realise how good we have things in New Zealand. We bitch and moan about our tax, the health system, poverty — all those sorts of things — but then you head overseas and realise how much worse people have it in other countries. I came back with a different view on life. Was I here to make money and run businesses successfully or was I here to make a difference? That was a big question I really started to think about.

My goals started to change after we got back. I wasn't so interested in being successful and wealthy in business or being well known on social media anymore. What I wanted to do was use my skills and knowledge to make a difference in the lives of people I met or interacted with online.

We become self-employed because we want to run farms the way we want to run them. There's all this information that we've learned, and we want to put it into practice in our businesses. This makes it quite easy to fall into the trap of being a perfectionist and only doing things your way. It can take a while of doing that to get to a point where you can step back and get some perspective to help you realise doing everything your way isn't necessarily the way to go.

What that trip did for me was force me to hand over the reins to Guri and Detroit and put all my trust and faith in them. They both really proved themselves. I was stoked with the job the boys had done. Being given the opportunity to do the problem-solving and having the chance to fail gave both of them heaps of confidence, which was awesome.

I told Detroit that I'd flick him $1500 to go on an OE. I told him that he could head overseas for six weeks any time he wanted to and I'd pay for some of it. I just knew that he'd really benefit from the experience. I also told Guri that he should go home to see his wife, who was still living in northern India at that time, and bring her back home here with him. I can't imagine what it must have been like for them living apart for so long.

Courtney came back and started her new job. While we'd been in Ireland, she'd had an interview over Skype with Mark Horgan for an agribusiness job at Westpac. It was really weird because it was in the daytime here, but it was 10 at night over there. She was really tired, but she must have done okay because they hired her! She absolutely loved the job and the people she was working with, which was awesome for her.

Even though I missed having her around, I really liked that she would come home and tell me how her day went and I could tell her about my stink day or my awesome day. It was great to have her to bounce ideas about the farm off, now she wasn't working with me all the time. We still both ran the gym together, but in terms of business I was running the farm myself and she was doing her thing.

As well as the farm and the gym, I was still playing rugby. Not long after we got back from Europe, I was back on the paddock playing for Marist. In one game, I went to tackle a guy, who was about to do a clearance kick from behind the tryline. I might have been a little bit late, but I didn't get caught for it!

After I tackled him, I got up and ran towards the line-out. As I was running away, he chased me down and smacked me in the mouth with his forearm. I got to the front of the line-out and my mouthguard fell out. I bent down to grab it and saw that four of my front teeth were sitting inside it. At least I was wearing a mouthguard even if it didn't help much that day!

For a while after that, I had to use words instead of a whistle to work my dogs. Even once I got new teeth, it took a bit for me to work out how to use them to whistle the dogs. I never realised just how much I needed those front teeth. I used them to my advantage, though, because once I started getting a bit of media coverage for Farm 4 Life a team from TVNZ came down to record a segment about me for *Te Karere*. I took my teeth out and completely freaked out Irena Smith, who was interviewing me. That sure got me — and the site — some attention!

While I was busy balancing the farm, the gym, rugby and Farm 4 Life, I found out I would soon have a new role to add to my already long list. For a while after we got home, Courtney felt a bit off-colour. She thought it was just a bit of jetlag or adjusting to being back. But it went on for a while and she was off her food and things were tasting a bit different to her, so she started to wonder if it could have been something else.

She took a pregnancy test in May and — HELLO! — it came back positive. It was one of the most amazing days of my life, finding out that I was going to be a dad.

We hadn't planned on having a baby — yet. We'd decided that we'd do another overseas trip — this time to South America so we could go to Machu Picchu — in April 2020. Then, after we got back, we were planning on starting a family.

Ever since I was about 11 years old, I've always thought about having kids. Even back then I used to think about what I'd be like and the things I'd do with my kids when I was a dad. From a young age I used to write down my goals and hang them on my wall. One of them was that I wanted to have my first kid when I was 28, then the second one two years later, and the third one a year after that. It was such a weird thing for a young kid to be even thinking about, but I was a bit of a weird kid! (I didn't quite hit the first age goal, but I was 28 when our little kina was conceived — probably in a hotel in Paris . . .)

I reckon the reason I started thinking about it then was because that was when I started working with Ian and Lisa Jeffrey. I saw the way that their family operated, the opportunities that they were able to give their kids, and how having money gave them options about how they wanted to live. That really influenced me. Now my dream was going to come true in a bit over six months' time.

I always knew we were having a boy. Whenever I told anyone that, they were like 'Oh, righto, it's a fifty-fifty shot. You can't know either way.' I was so sure, I bet seven people $100 that it was going to be a boy — and none of them have paid me. If you're reading this, you know who you are, you bloody buggers . . .

When Court had the ultrasound, they told me it was a boy and they were a bit confused when I said, 'Yeah, I already knew that.'

Courtney didn't really want to know if the baby was a boy or a girl until it was born, and I tried really hard not to tell her, but eventually I couldn't help myself. I was so excited.

During our time overseas, Farm 4 Life really started taking off. While we were travelling, we met heaps of people and talked to them about what we were doing, which helped drive traffic to the page. From that, it got heaps of buy-in from people in Ireland, England, Scotland, Australia and the United States. Even now, the site's international audience still makes up about 60 per cent of my followers. (I've even had a couple of my Irish Facebook followers come and visit me at the farm. I gave them a tour and promised them a tasty local treat. It turns out kina is a taste they haven't yet acquired over in Ireland!)

For people in the dairy industry overseas, it gives them some insight into how people overseas do things — especially those who have only small herds and find it hard to believe that we can milk more than 500 cows a day.

The audience is anyone who wants to watch, and my goal is to educate people about the dairy industry. I'm producing videos that I wish had been around when I was young. I think it would have fast-tracked my career and got a lot of my friends into dairy farming. Some people might find some of the stuff a bit challenging, but it's not a career for the fainthearted.

I like that people can watch it if they're lying in bed with their partner at 10 at night or if they're out getting the cows in at 3.30 in the morning. It's available to anyone who wants to see it whenever and however they want to see it.

One of the areas where I've found the page has really added value, even though I hadn't intended it to, is mental health. I talk openly about my troubles on the farm, and that's meant other people have felt comfortable coming out and talking about the things that worry them. It's helped to break down some of the isolation that people in the industry can feel.

SHIT DAYS HAPPEN SOMETIMES

If you are struggling with calving, have lost staff, are battling with the weather or any other crappy thing, just remember that everyone else has rough days sometimes, too. Even for farmers, social media can make us think that everyone else's lives are perfect and amazing — but they're not.

The more we tell each other when we're having shit days, the more normal they'll feel to all of us. If you are feeling that way, have a think about what you're eating, make sure you're hydrated, turn off your phone for a bit and, if you can, don't sleep with it in your room.

If you feel like nothing's going right or like everything that could go wrong has gone wrong, focus on the good things in your life no matter how small they might be.

Sleep deprivation, being under-hydrated and eating unheathy food to get through the day can affect the way you think.

There has been a bit of a downside to doing the videos. There's a heap of stuff online now where people are calling me a cow rapist and accusing me of stabbing live animals. (That one was after I put up a video about how to deal with severe bloat.) Sometimes I have to laugh even though it isn't funny. The people saying that stuff show how little they understand about what we do. Everything I do is best practice in the industry both for the cows and for the environment.

I guess those kinds of negative responses are what happens when you stick your head out. I understand that the people who say that sort of stuff are coming from a different place and have different beliefs and different ways of living. I know I'm doing what I'm doing for the right reasons, so it's pretty easy to just brush off those sorts of things.

That sense of knowing that I was doing things for the right reasons got thrown into focus when my Chiefs-loving mate from Midlands, Blair Vining, died. It was so sad. He would never have known the impact he'd had on my life.

Blair had been diagnosed with terminal cancer in October 2018 when he was only 39. I was absolutely shocked when I found out.

When they told him he was sick, he then had to wait for six weeks before he could get an appointment to talk to anyone about treatment because the health board had already spent too much that year. It would be fair to say Blair was pretty pissed off about that.

In the year after his diagnosis, he and his wife Missy put all their energy into lobbying the government to set up a national cancer agency so other people didn't have to go through what they were now going through. Blair reckoned it was because if anything happened to Missy or their daughters, he didn't want them having to wait that long for treatment.

Even with everything they were going through, they both stayed really positive. It was unreal. Eventually, their work paid off and the government agreed to establish a Cancer Control Agency and put a Cancer Action Plan in place. He and Missy also set up a sports foundation and started work on getting a charity hospital set up in Invercargill.

When Blair died, it really got me thinking about a few things in my life. He really made me realise that one person could make a huge difference, and if there were more people like him out there the world would be a better place. If everyone found out they were going to die, what life would they live? I'm pretty sure it wouldn't be the life they lived today.

I decided I didn't want to wait until I was dying to really start making a difference. What Blair did with his anger and frustration inspired me to add a lot more value in the things that I was passionate about. Passion is key. If you're passionate about anything in life, you'll draw people to you. They want to hear you because you have a different tone in your voice. You're a lot more excited, you're a lot more engaging, you're coming from a different place. There's no incentive for you other than to share your passion. That became so obvious with Blair when he was campaigning for change. He was so passionate about getting help for other people that everyone who heard him stopped to listen.

Blair's death also helped me come to the realisation that as I've become older I've taken on responsibilities and I want to do everything I can to help people, but sometimes people can take me for granted. My goal is to help people get into farming and to get out of bad situations in life.

A lot of people want a piece of me, but when Blair died I realised that I was saying yes so much that I was losing focus on what I really wanted to achieve. I had to take control of things and start saying no to people.

It's really hard to say no to people who want you to help them out, but once I got used to doing it I started to feel a lot better about the fact I was keeping on track with what I want to do and I wasn't getting distracted by what other people wanted me to do.

Soon I had another reason to say no to doing stuff I didn't want to do. In late November, our gorgeous wee kina Tekauenga was born. It was unreal. It was so cool I wished I'd had a kid 10 years ago!

Me and Courtney couldn't wait to have the baby, but then when he was coming we had no idea if we were ready or not. Once he was here, we just had to be ready. Like all first-time parents, we learned a lot really quickly.

One of the most crack-up things for me — and Court will probably kill me for saying this — was when, on about day three, they put Court on this machine to help her express some milk. It made me feel a little bit like I was at work . . . She wasn't too happy when I told her I didn't think I'd take her on as a top milk-producer each year! Jokes . . . She is an incredible mum and she coped with the whole experience amazingly.

Uncle and Auntie were super excited to have their first grandchild. Kelly's daughters are in Australia and they haven't had kids, and Peata's son up north hasn't had kids yet either. They were over the moon when Tekauenga was born. Now they're like, 'Hurry up and have more!'

DISCRIMINATION IN THE WORKPLACE

Ever since I was at school in Whakamārama I've always had heaps of close female friends. Those girls I went to school with there are all like my sisters now and they are awesome as women. My darling Courtney is a weightlifter and she's a boss. She's stronger and more powerful than me, and I've always known it. After seeing her have a baby, I have so much respect for every mum out there, but quite apart from that I have heaps of respect for every woman out there.

The dairy industry is really dominated by men and can be a tough place for women to be sometimes. I hear lots of stories about men hassling women in the sheds because they're tentative about putting cups on cows that are kicking out a bit while the men will just charge in there and use their brute strength to get the cups on. The reality is the gentle approach is far better for the cow, but a lot of farmers are so focused on getting milking done fast that they turn it into a male/female issue. This can lead to women in the industry not being given a chance to prove themselves or not being given the opportunity to speak up for themselves. This has led to many of them leaving the industry, which I reckon is a real loss for all of us.

The only way anyone can learn to do something is by being encouraged to do it. They don't have someone come and go, 'You're not doing that right. Get out of the way.' At one point in our lives, we all got taught to do the jobs we do, so I really urge everyone to take the time to give other people a chance and to encourage them to give things a go and learn.

A lot of farmers have grown up in a really male-dominated environment and they might not even realise that what they're doing is not working. Thankfully, that's changing, and we live in a different era now. If you're a man and you catch yourself talking or thinking like that, try to catch yourself and change your actions.

If this is happening to you at your work, call a meeting and explain what's happening to you, tell the boss that you're there to learn and you want to understand, and that if you're not doing something 'correctly' then ask someone to show you how it's done.

If you're in a space where you're dealing with negative men being dickheads, I really encourage you to stand up and speak out for yourselves. I recommend talking to people in the industry you admire and who have skills and knowledge they'll share with you.

At the start of each year, our teams set goals that we then look at in our weekly meetings. In those meetings, I encourage everyone — regardless of age, gender or ethnicity — to have a go at new things and learn new skills. This gives them the opportunity to have some control over their mahi.

If I see or hear anyone treating women badly in the industry or in the community then I will always call it out. Sometimes it can feel a bit awkward, but I know I don't feel anywhere near as awkward as the wāhine who are the target of that bad treatment.

15.

BUILDING THE HUB

How do you motivate somebody to play a sport when they can't score a goal? How do you tell your kid 'I want you to play basketball, but you're not going to shoot goals'? How do you tell them you want them to become an All Black without telling them about scoring tries? That's how people come to the dairy industry — they don't get shown about scoring the tries and shooting the goals or even what the hoops or trylines look like.

In the dairy industry it's like 'Get a job' and that's that. There's no guidance around what the process looks like for people outside the industry. It's all right once you're in it — you can find it — but if you're not in it, or you're not in it with people who want to share what they know, then you're stuffed.

As an industry, we don't sell the story of what a career as a farmer looks like, so we don't have a lot of people wanting to become dairy farmers or sharemilkers. Kids see an All Black and they want to do that. They know they have to work their way up, they know the path, but with farming they have no idea.

You can't just play club rugby and then be an All Black. It does happen every now and then, but not very often. It takes a lot of training, a lot of failure, setbacks like not making teams, not playing well or getting injured, being in the right place at the right time. Everyone knows that.

The process is the same with the dairy industry. People see the flash truck and the new car and they say 'Ohh, they own a farm!' But they don't realise it takes a lot of work to get to that truck, car and/or farm. Part of the problem is that the process of going from working on a farm to owning one isn't clear.

It's about getting the small wins. I sometimes picture the farming industry as like a game of basketball. It's not like soccer where you get a couple of goals and then you win. Instead you've got to get, say, 100 points before you win. You've got to chip away over time to set yourself up to win.

We don't celebrate every goal that we shoot along the way — that's why it's hard. What are the celebrations along the way? There aren't any. If you're lucky you might get into a manager's role or you might get a pay rise, but there are very long gaps in between. As a young person coming in, it's important to know what the process is and to celebrate the wins you do have.

There are a few different routes you can take to study for qualifications. Lincoln University, obviously, but even if you do that you still need to experience the practical side. Or you can go down the route that I chose. I wanted to earn money as I didn't have the funds to study full-time. That was the best for me — earning and learning through Primary ITO.

That was great, but I could see an opportunity to educate people about the industry at a way younger age, to get people into it, to sell the vision to kids from the age of, say, 11, 12 or 13. I'd had that happen to me. At the age of 11 I'd been influenced about going dairy farming, and then I got a job helping out on a dairy farm on the weekends when I was about 12.

There's a learning barrier stopping people from getting into dairy farming. You can't get into farming unless you have a job, but how do you get a job if you're not qualified or you don't have any experience? You can't get a job unless you have experience and you can't get experience unless you have a job. That's something that really needs to be addressed.

HUSTLING

Believe it or not, everybody has a little bit of hustle in them. Whether it means getting to December and your undies have got holes in them, so you end up turning them around and using the front for the back, that is called a hustle.

Learning to hustle is life. For me, hustling was rearing calves every year to make sure me and my partner could go on holiday no matter how much debt we had, no matter how much stress we were under. I reared these calves and none of that money was already accounted for, so we used it to take much-needed breaks away from the farm.

Hustle is vital, whether it be for growth, to get ahead, to step back from what you're currently doing, or to improve in areas where you're showing weakness.

Then there are the off-farm hustles. For me, they've been running my gym and starting up Farm 4 Life. None of that comes from standing in front of cows' arses 24/7. It comes from stepping back and letting your staff take responsibility, and from seeing what opportunities are out there not only at the farm gate but also in your community.

Whether hustling is about helping out the poor, running the barbecues for your school fundraiser, whatever it is, go out and do it, as it will add value not only to your life but to your community.

Over the years, I've had to teach the same things to every young person I've employed. I thought to myself, 'How long am I going to be farming for?' Another 35 years at least. In pretty much every one of those years, I'll probably employ a young person for a role on the farm. That's at least another 35 times I'll have to teach those very same things to a new person. How inefficient is that? We're in the twenty-first century here and we are still relying on one-to-one communication for basic knowledge-sharing on the farm.

Add to that the fact that the way I will teach them to do something will be completely different from the way the neighbour will teach them to do it, and things get even less efficient. Once I worked all this out, I started tossing ideas around about how to make this all more streamlined. I thought about organising training courses, where we'd get people to come away for a week and we'd teach them about farming. But that still only raised the numbers from one-to-one up to one-to-30.

I also thought about the mentoring role where you'd have a big brother system and match new recruits into the industry with more experienced workers, who they could go to for advice and support.

Then I thought about how everyone uses social media. I wasn't brought up with a computer — I didn't have one until I was 16. While all my friends were on Bebo and playing PlayStation, I was out building huts and stuff. I didn't really understand the whole social media thing until I had enough money to buy my own computer in my late teens.

All around me, everyone was using YouTube — young people, old people, even me. I did a bit of research and realised that you could easily measure how many views your videos get on YouTube. I realised that if I used a video platform, it wouldn't be one-to-one, or even one-to-30, but one-to-30,000. That there is efficient.

The other cool thing about it is that anyone can go back and re-watch a video, which means they don't have to piss any bastard off by saying, 'Oh, can you show me how to do that again? I forgot.'

I realised that a video platform would make knowledge-sharing way more efficient while also being way more entertaining and way more engaging.

For the content to have value for farmers, who were going to be employing these young people, I had to work out a way for their engagement with the videos to be measurable. The people using the site needed to be able to leverage off the time and effort they'd put into learning into getting a job. Not a farmer out there is going to be persuaded to hire someone who tells them they know what they're doing because they've watched videos on YouTube!

On the farm, I was meeting vets, bank managers, accountants, who'd been in the job for years. I've got the gift of the gab, so I would tell them I wanted to pick their brains and they would tell me whatever I wanted to learn. I was lucky to be in the position where I could ask them, and they were happy to help. I was also happy to stop them and ask them to break concepts down if I didn't understand them. I was confident to be a dumb arse in front of the person I was talking to.

I also knew that I had credibility in the industry because I was still farming, so I was relevant to the people who were farming. I was still young enough to be relatable to the young people coming through, and I had access to all of this incredible knowledge through my networks.

I realised that I had all these unique ingredients lining up together and that there was this huge opportunity in front of me if I could do it right.

For a while, this big plan went on hold while I focused on my family, the farm, the gym and building up the Farm 4 Life Facebook page. When I'd first set up the Facebook page, it was as a first step towards creating a knowledge hub. The plan was always to build the profile on social media and build the trust of the people who followed me, while bridging the gap between people who weren't farmers and people who were. It was also a good chance to help out people who were buying small lifestyle blocks and didn't really know what they were doing.

I'd been following this guy called Robett Hollis on social media for a couple of years. I really liked how real he was, and I really admired his business brain and his out-of-the box way of thinking.

I love his videos, and every time I see one he teaches me a lot. I really rate the guy.

Robett is Ngāti Porou and was born just out of Dargaville. When he was younger, he was a professional snowboarder, and he even won a silver medal at the World Championships. Since then, he's started up a heap of companies, including one called Frontside, which was a video production agency that got sold to Saatchi and Saatchi.

After selling his companies, Robett decided to focus on inspiring people to succeed and on trying to destroy New Zealand's tall-poppy syndrome. He reckons one of his biggest drivers is seeing other people win. He's next level.

Because I was thinking about growing this new business, I decided to get in touch with him in early October 2019. I thought he probably got thousands of emails every day so I thought, 'F— it, I'm going to do a video.' I reckoned a video should jump out a little bit and give me more of a chance at catching his attention.

I stuck my phone up on the visor of the tractor, did my hair nice, cleaned the cow shit off my face and made a video on Vimeo.

'Chur, Robett, my name's Tangaroa Walker. I'm a farmer down here in Southland and I can see that there's a huge opportunity in my industry to really uplift the amount of knowledge that's being shared from the top all the way down to the bottom in the industry. I've got this platform with nearly 100,000 followers, but I want to use that to bootstrap this other project I want to develop. I don't know how to do it, and I'm not sure if it's going to be any good. I'm really just after some help, bro.'

I sent him a link to the video through Facebook Messenger. A few hours later, I checked, and I could see that he'd seen my message. Then I checked the view counter on Vimeo and it had gone up — by one. He'd seen it!

A couple of days later, I got a message from him saying he'd loved the video and he'd be in touch soon. Then he invited me to the Powermoves Retreat he was holding up in Queenstown. I was absolutely blown away. I couldn't believe I was going to have the chance to go and spend a couple of days with Robett.

He'd handpicked about 80 people from around New Zealand to meet at the Hilton in Queenstown. Talk about high-rollers! Once I'd registered, I got sent a list of everyone who was going to be there. Courtney got hold of it and she was like 'Hinemoa Elder's going to be there!' I thought that was pretty mean.

NETWORKING

Networking is a huge one. It comes from getting out of the cowshed, out your front gate, into the community and meeting people. When someone comes up your driveway, whether it's the vet, the fertiliser truck driver, the soil rep, whoever it is, get off your arse and make time to go and see them. Talk to them and build relationships because relationships bring opportunities, and opportunities are what give you growth, and growth gives you value and money.

If you're driving past on a bike and you see someone who has just come onto your farm and has knowledge and value to add, go and ask them what they're up to, ask them where they're from, show some interest in what they do and they'll remember you and be there to help you in the future.

Get off the farm and meet people who inspire you. It doesn't matter if they're in different professions, get out there and meet them — you'll learn from them in ways you never knew you needed to learn. They'll also be there to give you advice when you need it.

I drove up to Queenstown and sat in the foyer waiting for everyone to arrive. I pretended to be on my phone, but really I was looking at who was coming in for the retreat. I saw Nathan McCullum, the cricketer, come in. Unreal! I thought he must just have been staying there, but it turned out he was there for Powermoves as well. Awesome. Then I saw Monty Betham walk in. I couldn't believe it!

Robett came in surrounded by this big bunch of people. I couldn't believe that it had gone from me sitting in the tractor, covered in cow shit, doing a video, to being here in the flashest hotel in Queenstown, with Monty Betham, who's like my bloody idol on the league field, and waiting to talk to Robett Hollis. Then Hinemoa Elder came in, then Hana Tapiata, then Charmaine Ngarimu . . . it was just completely unreal.

The whole vibe and energy of the group was completely infectious. I felt like I was finally surrounded by people who were like me.

People always ask me how I can work late at night and be up so early in the morning. I do it because I love what I do, and at the retreat I was surrounded by people who had a similar passion for what they do. You get 80 people like that in a closed room sharing yarns, it was awesome.

I was a bit starstruck when I finally met Robett. It's that weird thing when you know people through social media, so you feel like you know them really well, but you actually don't know them at all. Then when they're standing talking to you, it's a bit hard to process.

About 10 minutes later, he says to me, 'Do you mind doing a speech, bro?'

'To who?'

'To us — the crew.'

'What am I going to talk about?'

'Talk about passion and attitude, bro.'

'Okay, yeah, I can talk about that, sweet.'

I got up to talk right after Lance O'Sullivan, the doctor from up north who was the New Zealander of the Year back in 2014. No pressure, eh?

I couldn't believe that I, dumb-arse Tangaroa Walker who couldn't spell properly, was about to make a speech to some of the best and brightest people in the whole of Aotearoa. It felt like proof that if you shine your light strongly enough the stars will come down and grab you.

I really appreciated that Robett asked me to speak because, by doing that, he allowed everyone else there to learn about who I was. If he hadn't done that, I would have been just another person sitting in one of the seats and I would have had to go and introduce myself to everyone and tell my story over and over again.

Instead, I got to stand up on a platform and talk about my passion for farming and how I wanted to help people into the industry and teach people about how awesome farming is.

That weekend, I learned so much. It was so wicked to see what can happen when you bring together a group of people who trust each other and come with open hearts. Everyone was talking about their lives, their struggles, their wins and what they're passionate about.

After that weekend, it was solidified in my mind that I was going to charge 100 per cent at creating the Hub. I felt empowered and all of the shackles I had around me about creating it and all the insecurities I had about failure had been wiped out.

I came back from the retreat and set up a company called Farm 4 Life with me and Courtney as 50/50 shareholders.

A while before that I'd started branding up some merchandise for my team, and people started asking where they could buy it, so I decided to get some made. I approached a company in Pahīatua called Betacraft, who make workwear and wet-weather gear. I asked them if they would put my logo on their garments and sell them through their online store. They agreed to make me a brand ambassador for the company and to pay me a commission on the products that they sold with the Farm 4 Life branding on them. We've branched out heaps and we've now got about 40 different products in our store, and we're still working closely with Betacraft to develop new ones. It always blows me away when I hear about people ordering our gear from all over the world. It's mean-as to think of people living in Scotland or America wearing Farm 4 Life gear.

I'd been selling this merch on our online store and through Betacraft and I wondered if there were other companies that I might be able to work as a brand ambassador for. I talked to

Jamie McKenzie, who worked for Farmlands, and asked him if he thought I could get brand sponsorship for a new truck.

He suggested I talk to the guys at GWD Toyota, which I did. They were keen to jump on board as a sponsor for the new online video content we were going to be uploading for what we called Mask Off Monday, where I'd talk to specialists about mental and physical health issues. Dr Hinemoa Elder agreed to be the mental health expert on my Mask Off Monday videos, and my own GP Dr David Sarshalom helped out on the physical health side.

After that I approached Levno, who make our milk- and fuel-monitoring systems, and they were keen as to jump on board as a sponsor. All of a sudden, I realised how much potential there was for our new learning platform — Farm 4 Life Hub.

As well as giving them sponsorship rights to some of our content, I also did keynote speeches for the companies that supported us, which helped to further cement the awesome relationships I had with them.

Jamie McKenzie was a huge mentor around the project. He's very organised, disciplined and punctual, which I'm not, so he was a big help to me. He approached Ngāi Tahu on my behalf to see if they might be able to help me get some business advice. They agreed to give me some funding and put me in touch with John Schol, who was the CEO of local accounting firm, Malloch McClean.

I had a meeting with John about goal-setting. We spent three hours in the Malloch McClean boardroom as I unloaded my brain onto about 300 whiteboards! We talked about all the different things I wanted to put in the videos, how I wanted to create the Hub as a library of learning, and how I wanted to make the learning measurable.

Eventually, we realised that the videos would be organised a bit like books in a library. For example, the book might be about calving, then there might be a chapter about breach calves, then the pages could be broken down into the warning signs of breach calves, the equipment you're going to need during a breach birth, and so on.

John really focused in on how the Hub would be funded, and we decided on a subscription model where people would pay a set amount every year to be given access to all of the content, as well as discounts with the suppliers that we partnered with.

In terms of creating content, I knew I could call on my existing network of industry experts and also leverage off Farm 4 Life when I got in touch with other people who I wanted to talk with on specialist subjects.

That meeting was really the birth of the Hub concept, but there was still a long way to go before the concept became a reality.

SWITCHING OFF AND LOOKING AFTER YOURSELF

It's really hard to switch off when you work for yourself and you live where you work. Every day, there'll be someone driving up our driveway trying to sell us something, wanting a hand or needing information. It might be my day off, but if they see me sitting there, they're going to come and talk to me. There's no such thing as an out-of-office email for a farmer.

On days when I am working, there are so many jobs that need doing that can't be put off until tomorrow — especially when it comes to the animals. For example, once we've finished milking, the cows eat first. It doesn't matter what time it is, I don't go home for lunch until the cows have been shifted.

Then when I do go home for lunch, there'll be paperwork sitting there calling out for my attention, whether it's ACC and GST bills that need paying, bank statements that need checking or reports that need to be written. Sometimes I end up eating my lunch without even realising it because I'm so distracted by getting the paperwork done.

Even your lawns can make you feel guilty. Some people love getting home and pulling the mower out and mowing their lawns. It's their way of chilling out and relaxing. For a farmer, you look at your lawns and think, 'I've got to mow my lawns, then I've got to go over and mow the lawns at the other houses on the property and, once they're done, there's a couple of paddocks that need mowing before it rains...' There's nothing relaxing about that.

A lot of farmers don't do enough for themselves, and when they do decide to do something for themselves they often end up looking for validation that it's okay to take time out because they feel like they don't deserve it.

Of course, there are changes you can make to try to give yourself a bit more time out — not using your kitchen table as an office, getting off the farm on your days off, that sort of thing — but I think the most important change you can make is just becoming aware of it when you're feeling like you're not doing enough for yourself. Once you are aware of feeling that way, it's much easier to find ways of doing things for yourself like going to the gym, taking your kids to the park, catching up with friends, heading out fishing or whatever it is that makes you happy.

It's really easy for the farm to take over your whole life and before you know it you can get really down about it. Once you get to that point, it can be really hard to talk about it because it can feel like nobody else understands what you're going through. To anyone in that situation, I say there are plenty of people out there who do understand and who can help you. A lot of people have got support from the community on the Farm 4 Life page and another good place to start is your local Rural Support Trust (0800 787 254).

SEVEN FOR SEVEN

Another great way of gaining a new perspective on your life is thinking about where you are, how you got there and the people you have been grateful for along the way. One way I've discovered of doing that is something I call Seven for Seven.

Every seven days, I put aside seven minutes to phone someone who has had a positive impact on my life and thank them. I approach it as if it's the last seven minutes I'll ever get to speak to them, so I tell them exactly what they have meant to me. That might sound a bit weird, but if you go into the call with that mindset, it really focuses you on what you want to tell them. There are no rules and you're not worried about them judging you. The emotions really start to flow, and you can tell them how much you love and appreciate them.

I think people are scared of having conversations like this

because they worry about the other person's reaction. I do it every Monday night and I tell the person I'm talking to that I'm doing this thing called Seven for Seven where I call someone who I appreciate and thank them for what they've done in my life. As soon as you tell them that, the person just shuts up and they're all ears, they really want to hear what you want to say.

Every time I've made one of these calls, I've made not just the person's day but their month. It's been really cool to get these super positive reactions from people I really respect and appreciate.

16.

EVERYTHING CHANGES

I always had this idea that once Farm 4 Life hit 100,000 followers I'd really crack into setting up the learning platform from scratch. We hit the magic number at the start of 2020, so it was time for me to put my focus on creating a YouTube-like platform just for farming.

I wanted the site to be completely separate from the Facebook page, and I wanted it to be a place where people could learn about processes on the farm, but also where we could share knowledge about products that are used in the industry for the good of the farmer. I wanted it to be a place where we could share in-depth knowledge or new technology but also to talk about the problems that we experience on the farm and how to mitigate them.

I knew that if I was going to dedicate more time to developing the Hub, it would mean spending a little bit less time on the farm. It was really important to me that I be able to retain the staff I had working for me. I wanted to invest my money into the team culture and wipe out the worry that Guri and Detroit might have about losing their jobs every year when the end of the season came around. It's not uncommon for entire teams to change and head off in different directions each year, and that was the last thing I wanted to happen as Detroit, Guri and me made an awesome team.

I had a meeting with the boys where I said to Guri, 'I want to work with you forever.' And I said to Detroit, 'I really want you to align your attitude with mine and Guri's, and that will make us a perfect team.'

I let them know that if, together, we created a success of this farm, then I could afford to keep everybody together for as long as they wanted to be here. That kind of job security is pretty unusual in the industry and the boys bought into it straight away. We have an unreal team, we're all here for each other, and if anything goes wrong, we just help each other out and fix it.

LOOK AFTER YOUR STAFF

When you start out in a new job — especially in dairy farming — you'll be keen to get into heaps of cool new stuff, then reality will smack you in the face. While you might have been thinking about getting to spend all day outside and having a mean quad bike to do skids all around the paddocks, you'll soon get smacked with 4.30 am starts and 11/3 rosters that you never saw coming.

Some employers think about their staff as if they're a machine like a motorbike or a quad bike. You put some petrol in the machine and then expect it to get you from A to B and back to A again. That's all good, but did you check the tyre pressure, the coolant, the oil, or the water in the radiator? All of those things need to be checked often to get you from A to B all the time. It might be all good for a while, but you have to keep checking.

You have to get that vehicle serviced every six months or every so many kilometres. I did some sums for my truck, and I get it serviced after every 10,000 kilometres, which is after nearly 100 hours of driving. As an employee, if you're doing an 11/3 roster, you're working 11 days straight with three days off, which means you're working 132 hours every interval before you get some days off. If as an employer, we want to treat our staff like vehicles, then they're doing the equivalent of your vehicle doing 13,000 kilometres before it gets a service. And do you know what happens when a vehicle goes past 10,000 kilometres? The oil starts going black, it starts running really badly, it starts misfiring and things start getting clogged up.

If you look after your vehicle better than your staff, you've got to expect things to start going wrong — expect them to be hard to start on a cold morning, expect them to start blowing out a bit of smoke, expect them to start clogging up a bit.

Put things into perspective when it comes to employing staff. Treat them as well as you'd treat your vehicle (but not if you trash your vehicles, eh?). It doesn't have to be about more money. Do you know how much a feed is worth for a young person out on the farm away from their family? Do you know how much a warm house can mean? Do you know how much difference it can make having someone listen? It's easy to make small changes that will get them on board with what you're trying to do. It just takes a change in perspective.

SHOW GRATITUDE

I hear a lot of bosses complaining about staff not being how they used to be, that they're not resilient and they can't do long hours and all the rest of it. I reckon that's a mentality thing. Staff might have changed, but as employers we've changed, too.

When I first came into the industry I had awesome bosses who showed their appreciation by the way they fed me and the way they looked after me.

You might be paying your staff fairly, but can you do extra, can you do better? As bosses, we all want our staff to do more and to do it better.

We all expect a really high standard from our staff, but we need to deliver on our end, too. We need to push the boundaries of helping them out in ways that aren't financial. Think about dropping off food or bringing them in for dinner when they're working long hours and late nights. The same goes for contractors when they're working flat-stick on your farm. Always remember, a beer goes a long way when it comes to working a long day, say sitting 13 or 14 hours on a tractor.

> For workers, bosses are trying to meet your needs and requirements. If we can work together, we can make it an awesome farm to turn up and work at as a boss, as a worker, as a contractor or as a visitor.

———————

With the on-farm team all locked in, it was time for me to get to work on setting up the Hub. Then something we had no control over changed everything.

One of my best mates Tukes was living in Spain when the whole Covid thing started to hit really hard. In early March, he called up and told me what was happening over there and how it was beginning to look really bad.

I said, 'You might want to get home, bro.'

He said, 'Nah, it's all right. I don't reckon it's going to get too much worse.'

About a week later, he phoned again and said, 'Maybe I do need to come home.'

'Where are you going to stay, bro?'

He said he didn't know, so I told him that if push came to shove he could come and stay with us.

I wasn't sure how we were going to make it work, because the room where Tukes usually stayed was now Tekauenga's and the double bed in there had been replaced with a cot. Whatever, we'd sort it out when the time came.

Tukes booked his flights for the end of the month, then the government announced that the border was closed to all non-residents, and all returning New Zealanders had to self-isolate for two weeks after they got back. That's when things got a bit real.

I rang my mate Bull and asked him to come out and help me build a podcast studio out the back of the house, which could double as a bedroom for my mate.

Bull said, 'You might want to go grab all your timber then because everything's going to close.'

'What do you mean?'

'I reckon everything's going to close down real soon, so just go get what you need.'

I spent about three grand on ply, Pink Batts, timber, tools, nails, doors — you name it, I bought it. Bull came out and we put up the frames for the studio and we lined it. We didn't quite get it finished, but close enough.

It turned out Bull had been right. A couple of days later, on 23 March, the prime minister announced that the country would be going into alert level four in two days' time. That pretty much meant the whole country was in complete lockdown.

Everything happened really fast. I found it so weird how people panicked and went and bought a truckload of toilet paper (unless they were eating too many kina!) and food. As a food producer myself, it was clear that there wasn't going to be any problem with carrying on supplying people with milk, and we actually make toilet paper in New Zealand, so what was the point of it all?

I guess it pissed me off because I know what it's like to live week to week. The socio-economic chasm in our country was quite big before Covid-19 and I could see it was only going to get bigger. It really made me concerned for those families who can't afford to go out and buy three weeks' worth of food, who are living week to week, who are living in domestic violence situations and poverty. For them, it must have been really hard.

We had only a couple of days to shut down the gym completely, which was a really sad process. I guess the only upside of that for us was that it gave me and Court a little bit more time together as a family. Tekauenga was only about four months old when we went into level four lockdown.

Tuki made it back into the country just after we went into lockdown. I managed to organise for him to go and stay at my in-laws' rental property in Invercargill. It was empty because the dockers, who would usually be here at that time of year, hadn't been able to come. That meant I had two weeks to get the studio/ spare room set up for him, so he had somewhere to sleep once he'd finished his self-isolation.

On the Saturday night of the first weekend of lockdown, at about 1 am, I got a phone call. I thought, 'Who's died?' No good news ever comes at that time of night.

I got up and went out to the lounge. This voice on the other end of the phone was saying, 'Oh, hey, we've got this guy Keegan here. He's in the cells and he needs a bail address.'

I was really confused. 'What's that?'

'We need somewhere for him to stay for a little while.'

'Umm, how long?'

'It won't be long. He'll go to court soon and we'll sort it all out, but he really doesn't have any other options. If he can't come to your place then he's probably going to go to jail because he doesn't have anywhere else to go for lockdown.'

I still had no idea what was going on. 'Who? What's happened?'

The officer told me Keegan's full name and said he'd been up to mischief and needed somewhere to stay. It didn't help me much.

'Who gave you guys my number? I haven't seen him for about four years . . .'

He was a young guy who used to do weightlifting with Courtney when he was about 14. I guess he must have looked up to us a bit, because all these years on he'd put us down as his potential bail address.

I said, 'I've just had a baby, man. We've literally got an infant in my house. I don't want anyone coming to live with me. We've got enough going on at the moment.'

Tekauenga hadn't been sleeping very well. He was waking up two or three times a night. Court reckons part of that is because I don't know how to use my 'inside' voice and I was spending a lot of time on the phone sorting out business stuff. Sometimes, I would be up late at night working, then I'd be out getting the cows in at 3.30 in the morning, which would have been disruptive for him. As well as that, he must have found it weird that I was home much more than usual.

The cop said he understood, and he'd try to find somewhere else for Keegan to go. I felt a bit stink, but I just didn't see how we could help. There wasn't even a bed here for the young fella.

Five hours later, I was over at the shed when I got another phone call. They hadn't been able to find another bail address, and he urgently needed somewhere to stay if he was going to avoid being sent to prison.

I asked them if they could give me half an hour, then I came racing home to talk to Courtney.

She said, 'We're going to have to get him out here if he's going to stay out of jail.'

She never ceases to amaze me, my missus.

I called the cops back and they said we had to go and pick Keegan up. I headed into town and brought him back out to the farm where he ended up staying for the next five weeks. We cleaned out a tiny room that could just fit a mattress in it, so he had somewhere to sleep.

Luckily though, Keegan is a builder, so I told him we were going to get all his building gear and he was going to finish off building the studio for Tuki to sleep in, who was due out at the farm the following week. Keegan was pretty grateful to us and he happily sorted out the studio.

Once Tuki had finished his 14 days of self-isolation, he moved out to the farm to live with us and join our bubble as well. He had a broken leg and was really bored after spending two weeks in a granny flat in Invercargill, so he was really happy to be out with us at the farm.

So our home bubble ended up consisting of me, Court, Tekauenga, Keegan — a builder who was pretty much on home detention — and Tuki — a rugby player with a broken leg escaping Covid lockdown in Europe. It wasn't easy for me, but for my poor darling living in the house with all the boys and an infant baby it must have been an absolute nightmare.

Sometimes I wished I'd said no to them coming, but I couldn't have done it. There was no other way, really. I hate seeing anyone doing it tough, and if there's any way I can help I always will. It's the good old manaakitanga that my mum instilled in me from a young age.

Life on the farm during lockdown was pretty much the same as usual. The cows still got milked and the grass still grew. The only real difference was that our hygiene levels went through the roof. Our feed budgets were affected, though. Our cull cows would usually have been sent to the works around the time our grass growth rates drop. That was right in the middle of lockdown.

The freezing works were classed as an essential business, but because of social distancing they were running at low capacity while at the same time trying to get through high stock numbers. That made it hard for us to get rid of any animals. Even when we went back down to level one, there was still a backlog at the works because pretty much every farmer had excess stock they'd kept on their farms over the lockdown and it took a while before we could get a slot.

Leading into autumn and winter, we had to run the farm with extra numbers that we wouldn't usually have had. Coming into winter, our grass is at a bare minimum, so that's when we usually get rid of stock to the works. We couldn't do that, so we had to change the whole equation around feed for 50 days longer than we usually would have. Fifty days times 100 cows when you've got very little feed is tough in anyone's book.

As well as feeding the extra 100 cows, we also had to keep milking them past when we'd usually dry the herd off. We couldn't just stop milking them because we would have run into serious animal health issues, so instead of having our usual downtime to prepare for calving we carried on milking.

We had to open up our winter crops of fodder beet and kale earlier, and bring in more supplementary feeds like palm kernel and corn gluten. That meant we didn't carry as much feed through the winter as we normally would because we had to use it before we'd even dried the cows off. We also used up about 550 bales of baleage, which we would usually have used over winter.

Lockdown gave me something I really needed, though — time to focus on building the Hub. All of a sudden, I went from having had that one planning session with John through Ngāi Tahu, to talking on the phone with him for several hours almost every night for three weeks. I was working until two in the morning, then I'd be up at about five to go do the milking.

We ended up coming up with 3000 video concepts. Each individual video needed a time, a place, where it was going to fit into the Hub, who was going to sponsor it, where it was going to be shot, who was going to shoot it, and where they were coming from. We started off with a blank Excel spreadsheet and ended up with this huge beast of a document.

Once level four lockdown was over, the boys both left and we finally had our house to ourselves again. Shit, it was good! Not only that, but now I had a studio to do podcasts from and I'd laid down all the foundations for the next step of creating the Hub.

When things got back to normal — well, as normal as possible — I got in touch with Phil at Back9 Creative in Invercargill. I told him the vision and how I saw the final product, then he and his team set about building the Hub website. They were able to make it exactly how we wanted it in just four months.

One of the best things the Back9 team were able to do was make the content on the Hub shareable.

Now, if I'm away and Guri messages me to say 'There's a down cow and I can't get her up, I think she's got milk fever', I can go into the Hub on my phone, type in 'milk fever' and that chapter will come up. I can then share it with Guri and Detroit, so they can watch the video and see what they need to do. From my end, I can see when they've watched it. I don't even need to be there to show them what to do.

There's also a student version, where a tutor can use it to teach five or six classes every day. The tutor can drag and drop videos to her whole classroom, and the students will get a notification on their phone to tell them what they need to watch before the next class.

Then I had to go out and get businesses and specialists on board. I'd show it to businesses and say, 'We want you to occupy these 20 videos.'

They'd say, 'Oh my god, we can't shoot that much. We just can't do it.'

I'd say, 'Don't worry about it. Just give me permission and we'll do it. It'll take three hours of your time.'

I was blown away by how many of them agreed to come on board, and soon we had enough content to launch the first two phases of the Hub.

My goal for the Hub is to keep trucking along with three or four new videos a day for the next five years. I want to sexy up the merch so that we've got farmers looking a bit more swag out there. I want to bring on a female ambassador for Farm 4 Life. I want to try to broaden our knowledge around sheep and beef by bringing a section on them into the Hub, then maybe later bring on equine and marine sections.

Hopefully, I get to the point where we've got someone else who has as much passion for the dairy industry and a desire to add value for young people coming through as I do, so I can pass the reins on and get them to front all the videos and social media stuff. The whole reason I did this is because I thought someone needed to. If someone else had done it, I would have been happy to just keep on fishing and diving in my spare time.

In the meantime, I know I can be hard work. I don't know where I'm going to be because my plans are forever changing. I've got so many plates to spin at the same time between the family, the farm, the gym, the Facebook page and the Hub. As soon as one of those plates stops spinning, everything changes. Somehow, Courtney rolls with whatever is going on. She's just awesome. I can say to her, 'Babe, we're just going through a phase.'

Once Court's maternity leave was up, we had a big talk about whether or not she wanted to go back to work. Together, we decided that she would give up work so that she could spend more time with Tekauenga and be there for all those milestone moments in his life.

For me, as a dad, I'd rather she tells me about what he does during the day than have someone else tell me. We only have these years with him once, and I really want us to make the most of them. Although I can't wait for him to be walking. I can't wait to take him out on the farm and get him opening gates for me!

I really love the lifestyle I live, so I'll probably be a contract milker for the rest of my life. People often ask me if I want to own my own farm, but I'd rather invest in other things and diversify a bit like I have with the gym and the diving business. I also really enjoy helping people, so I'd like to do more of that. Having a family means I want to stay on the farm I'm on at the moment for as long as I can. I really enjoy working with the shareholders, and our consultants are doing a really good job.

TREASURE EVERY MINUTE

Everyone tells me I need to sleep more and slow down, but I've got so much I want to achieve and if I can do it in my twenties and thirties then why not? You never know how much time you have left, so get out there and do it now while you can.

Life is too short, and you have to do what you want to do. Don't wait around for someone to come and hold your hand. If you want something, just go and do it. Don't waste time doing stuff you feel obligated to do but that doesn't make you happy. Don't be a yes man (or woman).

There are people out there in relationships they shouldn't be in and they know they shouldn't be in. My advice to them is get out. You don't know how long you've got in this world, so you owe it to yourself to be happy. You might be working for employers who don't look after you and respect you. If you don't want to be there, leave, get out — you could die tomorrow.

I've lived my whole life at 100 miles an hour. I don't like sitting down and procrastinating. If I want to get stuff done, I'm pretty ruthless — I just go and do it. If you want to start a business, go and do it. If you want to sell your car and go travelling, go and do it. If you want to travel the world, go and do it.

You never know how much time you have. Tomorrow is not promised. As scary as it is, we live in a world of not knowing, so it puts everything into perspective to appreciate the time you do have and make the most of it.

Tell your friends that you love them. I tell my homies and

my boys I love them all the time. I tell my parents I love them. Sometimes I don't call people as much as I should, but when I do talk to them I tell them I love them. If you've got friends you haven't been in touch with for a while, ring them and tell them that you're sorry you haven't called for a while, but you think about them all the time.

Spend time with people who appreciate you. Check your priority list and make time available for the people who matter. Treasure every day, every minute, because you never know when it's going to be all up or when you, your partner or your family are going to be leaving.

ACKNOWLEDGEMENTS

I struggle to find time to ring people and tell them what they mean to me or to thank them for their contribution to my life. A lot of people have supported me, mentored me, employed me, dived with me, drunk with me, played rugby with me, worked with me, believed in me and loved me. To every one of them, I want to say thank you. You have made me who I am, and I do what I do because of you.

Tangaroa Walker is the much-loved face of Farm4Life and online education platform The Hub who runs a successful dairy farm in Southland, New Zealand. This also makes him a builder, plumber, electrician, engineer, truck driver, vet, accountant, employer, employee and primary ITO trainee. A graduate of the school of hard knocks, Tangaroa milked his first cow when he was 13 and went on to win the inaugural Ahuwhenua Young Māori Farmer Award and Southland Primary ITO Trainee of the Year Awards, earning six figures by the time he was 21. He has played rugby for Bay of Plenty and Southland Māori and runs the gym The Barracks in Invercargill. In his spare time (!) Tangaroa loves fishing, coaching and training and spending time with his whānau — wife Courtney and baby son Tekauenga.
Join The Hub at www.farm4life.co.nz